AMERICAN BUSINESS ABROAD

Origins and Development of the Multinational Corporation

This is a volume in the Arno Press collection

AMERICAN BUSINESS ABROAD

Origins and Development of the Multinational Corporation

See last pages of this volume for a complete list of titles

ORES AND INDUSTRY IN SOUTH AMERICA

H. FOSTER BAIN

AND

THOMAS THORNTON READ

ARNO PRESS
A New York Times Company
1976

Editorial Supervision: SHEILA MEHLMAN

Reprint Edition 1976 by Arno Press Inc.

Reprinted from a copy in
The Newark Public Library

AMERICAN BUSINESS ABROAD: Origins and
Development of the Multinational Corporation
ISBN for complete set: 0-405-09261-X
See last pages of this volume for titles.

Manufactured in the United States of America

Library of Congress Cataloging in Publication Data

Bain, Harry Foster, 1872-1948.
Ores and industry in South America.

(American business abroad)
Reprint of the 1934 ed. published by Harper, New
York, in series: Publications of the Council on Foreign
Relations.
Bibliography: p.
1. Mines and mineral resources--South America.
2. Mining industry and finance--South America.
3. Investments, Foreign--South America. I. Read,
Thomas Thornton, 1880-1947, joint author. II. Title.
III. Series. IV. Series: Council on Foreign
Relations. Publications.
TN34.B3 1976 338.2'098 76-4767
ISBN 0-405-09265-2

ORES AND INDUSTRY
IN SOUTH AMERICA

South America

ORES AND INDUSTRY IN SOUTH AMERICA

BY

H. FOSTER BAIN

MANAGING DIRECTOR, COPPER AND BRASS RESEARCH ASSOCIATION; FORMERLY DIRECTOR, U. S. BUREAU OF MINES; SECRETARY AMERICAN INSTITUTE OF MINING & METALLURGICAL ENGINEERS, ETC.

AND

THOMAS THORNTON READ

VINTON PROFESSOR OF MINING ENGINEERING, COLUMBIA UNIVERSITY; FORMER CHIEF OF INFORMATION SERVICE, U. S. BUREAU OF MINES; EDITOR "MINING & METALLURGY," ETC.

Published for the

COUNCIL ON FOREIGN RELATIONS

by

HARPER & BROTHERS PUBLISHERS

New York and London

1934

ORES AND INDUSTRY
IN SOUTH AMERICA

FIRST EDITION
I-I

COUNCIL ON FOREIGN RELATIONS

PREFACE

THIS book is the outgrowth of a series of discussions held by the Minerals Study group of the Council on Foreign Relations in 1931 and 1932. The group held in all eight meetings at which various phases of the relations of minerals to industrial development and international relations were presented by Messrs. H. Foster Bain, Sydney H. Ball, H. C. Bellinger, Clinton Bernard, Isaiah Bowman, Carlton P. Fuller, C. H. Haring, W. B. Heroy, Harold Hochschild, E. Holman, Frederic R. Kellogg, Charles K. Leith, Donald H. McLaughlin, J. F. Normano, Palmer E. Pierce, Thomas T. Read, George Rublee, Rufus S. Tucker, Arthur C. Veatch and C. W. Washburne.

The memoranda as presented were discussed informally but vigorously by all who attended, including from time to time Messrs. Arthur W. Allen, Amos L. Beaty, A. W. Benkert, Louis S. Cates, R. T. Cornell, Clinton H. Crane, C. V. Drew, J. Terry Duce, Stephen P. Duggan, Phanor J. Eder, Karl Eilers, H. B. Elliston, Dean Emery, Herbert Feis, A. Fleming, H. A. Guess, Charles P. Howland, Clarence E Hunter, H. W. Jervey, Sam A. Lewisohn, Walter H. Mallory, F. H. McKnight, Herman Metzger, Henry Mills, W. D. B. Motter, Jr., Vernon Munroe, Henry Kittredge Norton, William H. Osborn, Richardson Pratt, W. P Rawles, F. Bayard Rives, Leland R. Robinson, Edgar Rossin, Franz Schneider, Jr., William O. Scroggs, J. Edward Spurr,

G. B. Street, B. B. Thayer, Clare M. Torrey, Walter S. Tower, and Fraser Wilkins.

No stenographic report of the meetings was made since the purpose was to elicit the fullest and frankest expressions of opinion. In preparing this book, however, free use has been made of the memoranda which were submitted by those taking part.

The general purpose of the study was to secure a concise view of exactly what the mineral resources of the South American countries are, where they are situated, what part they have played in the history and development of each country, and to estimate, so far as that might prove feasible, the part they might be expected to play in the future. While it was recognized that minerals constitute but one of many factors in man's environment, it was held that they often are the critical factor in determining the extent and direction taken in industrialization. A self-contained nation must either be one which controls a wide variety and an adequate quantity of minerals, or one content with a very simple economy and a low standard of living. A nation not able or willing to face one or the other of these alternatives must be prepared to conquer and add to its domain territory which will yield the missing resources, or must enter into and strive to maintain commercial relations with others and thus become a factor in maintaining world peace. The significance of the individual minerals varies greatly in this connection and matters of quantity, quality, situation as regards transportation, altitude, and relations to each other are extremely important. The fuels are of first importance, iron ore ranks next and the others follow in turn. Even when the local supply is inadequate or of such character or situation as not to warrant final working up into industrial products on the spot, it may serve a high national purpose by affording through export as raw material the means of payment for products and facilities not otherwise to be acquired from abroad. It is to be granted that the will of the various peoples to industrialize or not to industrialize differs, and that with a particular people it may well change from time to time, but to any major program of industrialization minerals are essential and their occurrence in or absence from a country is a fact of nature beyond the will of man to change. It is believed, therefore, that exact knowledge of the extent and character of such occurrences and a broad knowledge of the implications of minerals in modern civilization are necessary to any sound interpretation of modern social problems. It was with this in view that the great continent to the south of us was taken as the field for one such study.

In the conferences of the group Mexico as well as the South American countries was considered but for the sake of unity South America only is reported on here. In preparing this book the authors have drawn freely upon the published literature as well as on their personal notes made while working in one or the other of the countries discussed. Nearly every South American country supports some sort of a geological survey or Bureau of Mines, several of them being of a high degree of excellence. Their publications have been extensively consulted and in repeated instances are cited in the text. In general, however, the effort has been made, where it has seemed desirable to indicate a source for further studies, to cite texts in English and by preference those easily accessible since this book is intended for those interested in social and economic studies rather than for the students of minerals. Three books in particular constitute veritable mines of information. These are: "The Mineral Deposits of South

America," by B. L. Miller and Joseph Singewald;[1] "South America; A Geography Reader," by Isaiah Bowman;[2] "South America," by Clarence F. Jones.[3] The authors' indebtedness to each of these is too large to have permitted citation from page to page. Miller and Singewald, in particular, have covered the whole subject of mineral occurrence in the region so carefully and have in each chapter cited so fully the outstanding reports and documents that it has seemed unnecessary to cumber this book with a bibliography. Instead, footnotes have been inserted where more recent reports or data from farther afield have bearing on the matter discussed. The following articles about minerals which have appeared in *Foreign Affairs* are of significance in this connection: H. Foster Bain, "World Mineral Production and Control," July, 1933; A. C. Bedford, "The World Oil Situation," March, 1923; Albert D. Brokaw, "Oil," October, 1927; Harry A. Curtis, "Fertilizers: The World Supply," March, 1924; Herbert B. Elliston, "The Silver Problem," April, 1931; James R. Finlay, "Copper," October, 1925; C. K. Leith, "The Political Control of Mineral Resources," July, 1925; "Exploitation and World Progress," October, 1927; "The Mineral Position of the Nations," October, 1930; Josiah Edward Spurr, "The World's Gold," April, 1925; "Steel-making Minerals," July, 1926; Walter S. Tower, "The Coal Question," September, 1923; A. C. Veatch, "Oil, Great Britain and the United States," July, 1931; Jacob Viner, "National Monopolies of Raw Materials," July, 1926; and H. T. Warshow, "Tin: An International Metal," April, 1927.

The authors also desire to acknowledge especially the

[1] McGraw-Hill Pub. Co., New York, 1919.
[2] Rand, McNally Co., New York, 1915.
[3] Henry Holt & Co., New York, 1930.

help of Messrs. M. Archila, J. M. Sobral, Eric Davies, J. Terry Duce, W. B. Heroy, W. D. B. Motter, Jr., C. E. Arnold, Sam A. Lewisohn, L. W. Strauss, A. M. Tweedy, and C. V. Drew, each of whom read and criticized one or more of the chapters covering areas with which he has especial familiarity. While, therefore, as always when one writes of a subject of which many have written before, our indebtedness to our predecessors is large and varied, we must exempt each, save where specifically cited, from responsibility for any specific statement in the text. It has been ours to select and arrange and, to the extent of our ability, to interpret.

Finally, lest some readers fail to perceive it, we would call attention to the fact that although our theme has necessarily made us discuss the importance of material things, we are not insensible to the profound truth in the observation of Robert Louis Stevenson, "The ground of a man's joy is often hard to hit. It may hinge at times upon a mere accessory . . . it may reside in the mysterious inwards of psychology . . . It has so little bond with externals . . . that it may even touch them not, and the man's true life, for which he consents to live, lie together in the field of fancy."

CONTENTS

ORES AND INDUSTRY
IN SOUTH AMERICA

CHAPTER I

THE PROBLEM

SOUTH AMERICA has been frequently called the Land of the Future. Wallace Thompson begins the first chapter of his recent "Greater America"[1] with this sentence: "For four hundred years, one of the mightiest reservoirs of wealth in human history has lain, virtually untouched, in the midst of the world." Many other expressions of faith in its undeveloped possibilities might be quoted. That it would, perhaps within the next century or in less time, be the scene of a repetition of the industrial triumphs of the past hundred years in Europe and North America has been fairly generally held. With its immense area, its varied topography and climate, its numerous resources and its sparse population, it has indeed seemed to be a land of opportunity. Its resources in coffee, wheat, cattle, nitrates, copper, and iron ore are well known to be of world rank. Others have seemed to await but organizing skill to place South America on an economic plane with the North Atlantic continents. In the people of South America are mingled the strains of the hardier, more resistant native peoples and the more venturesome and energetic members of the races of Europe. For four centuries this mingling has been in progress and the great outflowerings of various civilizations in the past have followed, if not been caused by, just such merging of the

[1] E. P. Dutton, New York, 1932.

talents of different races. So the world has rather taken for granted that history was to repeat itself on this southern continent.

But will it? Will South America become one of the great industrial regions of the world or continue, in the main, its present rôle as a provider of raw materials? Will circumstances or the inherent tendencies of its peoples conspire to defeat the widely held, but possibly poorly based, anticipations as to its future? Are the engineers, and their supposedly wicked associates the industrialists, to work their will there as in the North Atlantic and "apply the forces and energies of nature to the good of mankind" as they see it, thus building up a mountain of tangible goods for its peoples? Or is South America to be saved from the engineers and somehow to work out a radically different solution of the puzzle of living? Such questions have begun to be asked. Doubt has crept within the citadel, and whether it is in fact possible to unroll another magic carpet and transport the people of the south quickly into a land of abundant goods, whether if this could and should be done the ones who accomplished it would be hailed as benefactors or hissed as disturbers of the peace, is now less certain than at one time seemed, and the answer is of definite importance to the rest of us.

The relative power of peoples is a large factor in international relations. In an ideal world the strong and the weak would stand as full equals and share as such, but we are yet far from this ideal condition. The building of a powerful fleet by Germany was profoundly disturbing to Great Britain and a considerable factor in precipitating the World War. The creation and maintenance of a modern navy by Japan has certainly altered the situation in the Pacific. Are we likely to see the rise of a comparable naval force to the south of us? This is only one of the possible consequences of the complete industrialization of South America if Western Europe and North America are to be accepted as the models of attainment.

Industrial civilization, as worked out in the countries surrounding the North Atlantic, rests upon two major factors: the character and will of the peoples of the region and the resources at their command. The North American Indians had available for untold centuries all of the materials and forces that their successors have put to such intensive use, but they remained in the hunting and fishing stages of civilization. Only far to the south, in Mexico, Central, and South America had any Indians made much progress into the agricultural stage or had they organized effective governments which controlled large areas and millions of people. How far they had succeeded, prior to the coming of the Europeans, is sketched in later pages, but with rather unimportant exceptions both continents lay fallow prior to the coming of Columbus. It was the intrusion of another race that made North America what it is. It may be argued with reason that the white race was not a better race, but certainly it was one which had received certain mental stimuli that had not acted on the indigenes of America. There was a will to do which the native lacked and already a larger command had been acquired over the forces and materials of the earth, notably, as was of critical moment at the time, in the matter of making iron and gunpowder.

What any people have learned to command is evidently a large factor in what institutions they will develop. This seems to hold true to a large and as yet unmeasured degree despite the present quick and ready means of exchange of knowledge. With books and newspapers readily translated and transported, with cables, telegraphs, telephones, the radio and motion pictures universal, no people now is isolated. The knowledge of one is open to all and ideas pass quickly from one to another. Ideals are transferred more slowly. The will to accept is an important element and into this enter inheritance and race instincts which are of long, slow growth. One may know how another man lives and still, rightly or wrongly, prefer to continue his own style of living. The world now concedes this, at least in principle. That does not, however, alter the importance to each of us of knowing how our neighbors do live and whether we may expect them to continue to hold land in large areas and peonize their workmen—if that happens to be their style of living and happens to annoy us—or to all move into apartments and turn on the loud speakers in nice neighborly fashion. What, in other words, are they likely to want to do in the years to come?

Only the most uncertain answers, it must be conceded, can be made to such questions. The will of a people rests on many things and it changes, generally slowly but occasionally rapidly, in response to most unexpected events. One needs but to recall the truly revolutionary changes made in our political structure at Washington under the present Administration as a result of the bewilderment induced by the depression. The latter, however severe, is sure to prove but an incident in our whole economic history, but the political changes may well prove permanent, at least in part. He would be a brave prophet who would undertake to foretell what a people will attempt to do. It is a much easier, though still far from simple task, to point out what a people have to do with and actual accomplishment is ever measured by the means at hand as well as by the mentality of a people. It was formerly held a hardship to be forced to make bricks without straw. Modern science has circumvented that particular difficulty but there is still no known way to build brick walls without brick, nor is there any acceptable method to manufacture modern armament without steel.

In the industrialized states, where man has reached out and forced into subjection a greater number of the known forces of nature, wealth, either national or per capita, is not measured alone by the acres available or by the fertility of the soil. A very small and barren area, if it be the site of a mine, a water power, or a particularly good harbor, may make a contribution out of all proportion to its areal extent. It is this more intensive use of the materials and forces that marks the difference between the more highly industrialized nations and those living under simpler agricultural economies. It is because of this that the United States, with only 7 per cent of the inhabitants of the world, does roughly half its work and in recent years has forged rapidly forward in wealth. Back of each person in the United States is power, derived from fuels or falling water, equivalent to forty other men; men who work but do not eat. They produce but do not trench on the food supply. The net effect is an increased amount of goods to be divided and this mounting flood of wealth has grown much faster than has population.

Under an agricultural economy there is nothing to do with a surplus but to export it in exchange for other goods; if no foreign market is open, nothing much can be done about it. For the world as a whole food consumption increases only with population. Durable goods, on the other hand, especially those made from minerals, are subject to no such limitation. The total that can be consumed in any country is measured only by the wants of the people and the ingenuity of the manufacturer in meeting them. If the country needs no more railways, steel can be put into automobiles and when that market is saturated metal houses may follow. So long as developments are kept in balance, exchange can be kept up and both production and consumption can be increased. Foreign trade helps. It is the sensible easy way to bring about sound economic and political relationships so long as each party to the exchange offers something the other wants. But the big trade of industrial nations is internal and the possibilities of growth through finding and filling new wants at home is not restricted to the rate of increase in population.

It is this that makes mineral resources so important, for the durable goods are almost entirely made from minerals and in their making still other minerals are employed, as fuels, tools, machinery, or structures. It was the discovery of the utility of minerals and their utilization on a large scale which made possible, if it did not indeed cause, the flowering out of modern industrial civilization. In our own Colonial period wood was the all but universal fuel. Coal had been discovered but was as yet a curiosity and our largest mining industry was still to be created. Candles were depended upon for light. Very little iron was made; indeed, we used only seventeen pounds yearly per capita. Now we require more than fifty times as much. Lead was known, but the amount available was insignificant as judged by modern standards. Brass, made from zinc and copper, was so precious that individual buckles, pins and buttons were handed down in wills as heirlooms. Hundreds of minerals and mineral products now in wide daily use were then either unknown or regarded merely as natural curiosities. Conditions were much the same then throughout Europe. Elsewhere, even smaller quantities of metal satisfied the then known wants of the people.

The pinch seems first to have been felt as regards iron, and with even a slight increase in production so many forests were cut for charcoal that coal and coke had to be requisitioned. This was so revolutionary that less than a century ago Pennsylvania sent a commission to England to discover whether it was in fact true that coke made from coal could be used to smelt iron. As the demand for coal grew, pumps to unwater the mines had to be more powerful and so steam came into its first important use. Steam engines for power and for railways followed and the whole flood of modern machine industry welled up. It would lead too far from the subject to point out step by step how it all came about and to indicate even in outline how it has influenced the habits of life and the thoughts of the peoples of the world. It is sufficient to make clear that an abundant and varied supply of minerals lies at the very foundation of modern industry. Without them the latter would not be possible and if a new area is to be industrialized in the sense that word is usually employed, an adequate supply must be available.

Minerals may be classified loosely into fuels, metals, structural materials, and the raw materials for the chemical industries. The fuels are the energy givers. It is they which lengthen and strengthen the arm of man to do his will around the world. They increase his power and in a sense annihilate time and space. They alter the relative power of nations materially and so it comes about that China, with two and one-half times the population of the United States, has only one-ninth its physical strength; Italy, with the same population as France, has less than half its power, mechanically reckoned.

Not all power is derived from mineral fuels. Water power, wind power, animal power, slave power all are used. Not even all fuels are minerals, since wood, brush and other unchanged plant remains are commonly used. The larger part of the work of the world is done by burning the principal mineral fuels, coal, petroleum and natural gas. Of the total work done yearly in the world, about 6 per cent is done by animals, 12 per cent by human beings, 9 per cent by water power, 17 per cent by petroleum, and 56 per cent by coal. That done by wind is probably less than 1 per cent; its relative importance and that of animals as a source of power is declining. This is because the ratio to human work of the total so accomplished has changed but little over the centuries, while the relative amount of work done by water power has increased and that done by coal and petroleum has multiplied enormously. The last three make possible immense aggregations, or focalizations of power that have greatly transformed men's ways of living as well as made available to them greater quantities of the products of work.

Of the coal used in the United States more than a quarter is used by railroads, more than a fifth by the iron and steel industry, another fifth goes for domestic and non-specified uses, about a fifth is used in general manufacturing, less than 10 per cent by electric utilities, and less than 2 per cent by plants manufacturing gas. Comparing this with present needs in South America, several implications are evident. A good part of that continent lies between the Tropic of Cancer and the Tropic of Capricorn, so that there is relatively little necessity for domestic heating; even where high altitude requires it the European habit of economy in househeating which is followed makes the demand for fuel for this purpose relatively light. There is no iron industry of importance in South America, and since the region only produces 1 per cent of the world's supply of coal and consequently has correspondingly little to haul, it needs less railroad transportation, which reacts again to produce less demand for coal. Coal consumption in South America declined from 16 million tons in 1913 to 12 million in 1929. In attaining a realization of what this means, the following table will be helpful.

TABLE I

	Daily Output of Work—Millions of Horse-power Hours					Daily Output Hp.-Hr. per Capita
	Human	Coal	Petroleum	Water	Total	
Chile	1.5	2.0	3.8	1.0	8.3	1.90
Argentina	3.6	4.5	10.0	0.4	18.5	1.69
Peru	1.8	0.4	0.9	0.5	3.6	0.66
Brazil	13.0	3.4	3.0	6.0	25.4	0.63
United States	40.0	1,001.0	481.0	121.0	1,643.0	13.38

The basis upon which this table has been calculated is explained in the *American Economic Review* (March, 1933, p. 55). It appears from it that while the significance of water power is about the same in South America as it is in the United States (corresponding to somewhere between 5 and 10 per cent of the total energy output), the relation of coal and petroleum is reversed, since here about two-thirds of the total work is done with coal, while in South America it is petroleum which contributes the two-thirds. This, however, is not because petroleum consumption there is large but because coal consumption is small. In the United States the per capita output of work is 13.38 horse-power hours per day, or over seven times that of Chile, and in Europe the corresponding output is two or three times that of the most developed South American country.

This sets forth in quantitative measure the general observation that South America has as yet an agricultural and pastoral rather than an industrial economic status. Its people are engaged in the production of food and raw materials rather than in manufacturing enterprise. This, of course, reacts to limit their present need for mineral commodities in a variety of ways. Therefore, even though the per capita production of minerals is not large, it is still able to export minerals in considerable quantities. There is perhaps no other large region in the whole world from which so large a part of the mineral raw materials produced is exported. The individual countries, of course, differ markedly in this regard, as will appear from detailed consideration of their statistics.

Among the mineral industries of growing importance in South America is that of the production of petroleum. The coming of the internal-combustion engine transformed that industry from a kerosene-producing to a gasoline-producing industry, but while this solved the problem of what to do with the former by-product, gasoline, it created a new problem of what to do with the fraction of the crude that is heavier than kerosene. In the modern state of knowledge it is possible to "crack" this into gasoline, but only at a higher cost than "straight-run" distillation so long as more crude is available than immediate demand requires. The petroleum industry normally now has available for sale large quantities of heavy residue from distillation which, on the average, it has to sell in competition with coal. Since maritime freight rates are based on the cubic foot of space rather than on weight, and since a cubic foot of this heavy residue contains about 50 per cent more heat energy than a cubic foot of broken coal and in addition it is under most circumstances much more convenient to handle, it is clear that petroleum has another reason for regional importance in South America besides its relatively greater abundance there. Power plants there use heavy oil as fuel where in the presence of a normal supply of coal, the latter would be expected to be used.

Unlike coal, petroleum requires processing before it can be used and that, in turn, requires attaining some balance in the markets for the different products. The petroleum-refining industry is, however, usually operated on the basis of the gasoline demand; so much of the other products as can be marketed is produced, and the remainder sold as fuel oil. It is also not essential to complete the refining operation in the plant in which it is started, so that the petroleum industry is in some ways more flexible than the steel industry, which is next in qualitative and quantitative importance in modern civilization.

The general absence of coal in South American lends particular importance to any information bearing on the extent to which petroleum and natural gas are present. The particular situation as regards each country will be reviewed in the individual chapters. It will be sufficient to say here that while petroleum has been developed on both the Pacific and Atlantic coasts, the largest amount has so far come from fields along the east front of the Andes. According to geologists familiar with the region, there are indications favorable to the occurrence of petroleum in a belt extending from Venezuela to Patagonia. Surveys have been made and exploration has been undertaken at a number of points. No adequate data exist for estimating the possible or probable reserves. Indeed, the attempt to make such estimates for areas much better known has not been sufficiently successful to encourage geologists to continue to do so.

Garfias[2] has, however, recently presented some interesting figures on actual oil reserves which are of interest as indicating the present position of South American

[2] V. R. Garfias: "An Estimate of the World's Proven Oil Reserves," *Trans. A.I.M.E.*, vol. 103, p. 352, 1933.

countries as compared with the United States and Russia, the two leading producers. He believes there is a value in attempting to estimate the "proven" reserves—"the oil that yet remains underground in producing fields—and their logical extensions," despite the failure to estimate probable reserves usefully. He adds that it is necessary to remember that such estimates remain of value only during the limited time when the controlling factors remain unchanged.

ESTIMATED PROVEN OIL RESERVES
(Millions of Barrels)

Country	Proven Reserves	Production 1932	Total Production to Dec. 31, 1932	Years of Past Production
United States......	12,000	784	14,787	74
Russia............	3,000	154	2,883	72
Venezuela.........	2,000	121	761	16
Colombia..........	400	17	121	11
Mexico............	300	32	1,665	32
Peru..............	100	10	141	37
Argentina.........	100	13	95	25
Trinidad..........	90	10	84	24
Ecuador...........	10	1.7	7	16

It is to be anticipated that intensive exploration and favorable conditions for development will permit extensive additions to some or all of these reserves. It is to be regretted that data are not available for similar estimates of other mineral resources.

Few people realize that the world requires fifty times as much steel yearly as it does of any other metal. This is because steel, in addition to being cheaper than any other metal, is useful for many more purposes than the others. Principally, however, it is because men are now making things out of steel that originally they made of wood. The steel automobile and truck have supplanted the wooden buggy and wagon, the steel steam vessel has displaced the wooden sailing vessel, wire fences take the place of the rail fence, and above all, the steel frame building occupies the rôle of the frame building. In the case of the wire fence a small amount of steel takes the place of a larger quantity of wood, and in general a small steel member is as strong as a larger wooden one. But the effect, in nearly every instance, is to permit the building of much larger units; the thirty-story office building takes the place of the three-story frame building. This not only calls for large quantities of material, but calls for materials delivered in the required form, instead of being shaped on the job by skilled workmen. Even more important is the requirement that the delivered material must be of uniform high quality. The outstanding trend in the steel industry of recent years has been toward high quality and uniformity in its products.

Alloy steels are of so much public interest because of their special qualities that there is a tendency not to realize that the big bulk of the steel made is still of the quality known as "mild." Just as most flour is turned into bread, in spite of the interesting variety of cakes and pies a bakery displays, so most of the iron produced is converted into steel of good and uniform quality to sell at a low price. Sheets, bars, wire, rails, structural shapes for buildings, bridges, *etc.,* absorb most of the world's output and are of a quality that sells at a low price.

The equilibrium between the local use of semi-finished material and reliance on finished products is not easy to picture. No one would attempt to make pipe or rails, for example, in an area of small consumption; such products are transported thousands of miles because the economy of large-scale manufacture is necessary. Every industrial plant, on the other hand, is continually working up sheets, bars, and rods into things that are either of special design or are not needed in large enough quantity to order from a distance. Local industries develop where an adequate market exists. The situation parallels the one indicated above, where the production of bread is handled by big companies, local bakeries develop a small business for their special product, and the average householder keeps a little flour in stock for general use and to permit anything special to be made.

The steel industry is very different from the baking industry in the use which it makes of old material. Iron can be remelted and cast any number of times, and any industrial plant large enough to have its own foundry is always reworking its scrap material. In the steel industry quantities of old material are added to new iron in the steel-making furnace and it is possible, indeed, to operate wholly on old material. This means that steel-making is not necessarily limited to a region in which iron can be made from ore. In the United States, no new iron is made west of Utah but there are steel plants of considerable importance in California and Washington. The steel-making process can utilize a variety of fuel materials, instead of being dependent on a special type, as iron-making is; consequently steel-making is likely to develop in South America in advance of iron-making, but so far neither has attained any importance. Of the minerals accessory to steel-making, alloy metals such as manganese and vanadium are the only two of which deposits of major importance occur in South America. These minerals, of course, have to be exported, as there is no local use for them.

Practice in regard to some of the more important nonmetallic minerals has changed but little throughout the centuries. A brick house is still constructed by spreading mortar on a brick and placing, by hand, another on top of it, while the making of stone walls has been facilitated only to the extent that machinery is used in breaking and shaping the stone and setting it in place. In a modern building the outside brick wall does not even rest on the foundation, being carried on the steel frame. It contributes nothing to the strength of the structure, serving only to keep out the weather, and is consequently made as thin as practicable. The stone work in such a building is employed for its appearance and not to lend strength to the construction. For an important building stone of the desired color and texture may be brought halfway round the world, but stone is no more essential to a modern building than bird feathers are in the construction of women's hats. The mixture of crushed stone, sand, and cement, known as concrete, is, however, essential for economy in construction. The steel frame must be protected by it from the effects of fire for if steel is heated to 1000° F. it loses half its strength. A modern building is not "burned up" but collapses from the effect of the fire. Floors and walls are made of concrete and massive foundations are cast in single pieces. Cinders, from the combustion of coal, are used in place of crushed stone in concrete for floors and walls, where it is desirable to reduce the weight and great strength is not necessary. It is estimated that in the construction of Rockefeller Center, in New York, 88,000 tons of Portland cement, 245,000 tons of sand, 150,000 tons of cinders, and 42,400 tons of crushed stone were used, together with 125,000 tons of structural steel, 72,000 tons of brick and 43,000 tons of limestone.

The point of interest to us here is that sand and stone suitable for crushing can be obtained almost but not quite anywhere in the world, so they need never be transported any considerable distances, one of the few exceptions being in the vicinity of Buenos Aires. The raw materials of Portland cement, clay and limestone are widely available, though here again the great pampas of South America show a deficiency. But in making cement, after grinding and mixing, it has to be burned, and about three-quarters of a ton of fuel is needed per ton of product. It is again ground very fine and protected from moisture until it is used. To do all this properly and economically requires a considerable investment in plant, consequently it is an enterprise that can rarely be initiated in a small way. Brick, on the other hand, can be burned in small kilns. Consequently, as would be anticipated, South America is self-supporting as regards brick, clay, stone and sand, but mostly depends on the outside world for Portland cement and structural steel.

One of the important uses of limestone is as flux in the iron blast furnace, and in South America it has no such field. Another is in connection with beet-sugar manufacture, and here again the need for the mineral does not exist, while the important use of limestone in connection with all sorts of chemical industries is almost completely lacking. An important use of sand is in connection with glass manufacture and here also little market exists for it in South America because the industry that utilizes it is so little developed.

This general relation applies to a great number of non-metallic mineral commodities. Such things as magnesite, barytes, feldspar, fluorspar, talc, Fuller's earth and various others have little *raison d'être* for production in South America, for the industries that utilize them are either absent or are as yet only rudimentary. Only where deposits of exceptional quality or unusually favorable conditions for low-cost production and export exist would we expect to find them developed.

Copper, lead, and zinc, the three metals of which the world requires between one and two million tons each yearly, are of unequal interest in South America, for in the irregular distribution of their deposits around the world it so happens that South America has none of major importance for either lead or zinc. For copper it was, until the recent development of copper in Africa, long the continent of second importance as a producer, and Chile was from 1850 to 1880 the leading producing country of the world. Copper can be produced on a small scale if the ore is rich enough, but the major part of the world's supply is now derived from large low-grade deposits, and the crude copper produced by smelting operations must be sent to a refinery to be brought to a high degree of purity and to recover any gold, silver and other metals that may be associated with it. Copper is used in windings of dynamos and motors, in telephone wire and cable, in manufactures of brass, bronze and other alloys. As already indicated in the case of steel, a plant to produce these things cheaply and efficiently must have a large and wide market. So, even in the case of the one country in South America which is a large producer of refined copper, the metal is exported and manufactures of copper are imported.

Deposits of bauxite, the raw material from which aluminum is made, that are of suitable quality for efficient production of the metal are rather scarce throughout the world, and South America is fortunate in possessing important deposits which are described in detail elsewhere. But to make aluminum from its raw materials requires electric current produced at a low cost since the energy required is about twenty times as much as for iron. Consequently, it is typically produced at places where electric energy can be generated cheaply and where there is no more profitable use for it. Without venturing an opinion as to whether any water-power sites in South America meet this condition, the fact may be noted that they have not so far been developed except under circumstances where the energy can be marketed at a higher rate than an aluminum plant can stand. Consequently, South America's bauxite output is being exported and will probably continue to be during the life of the deposits.

Tin is a relatively valuable metal, ordinarily selling for about three times as much as copper, and its production from its ores does not require expensive plant. But it reaches the consumer in many fabricated forms, as containers for preserved fruits and vegetables, tobacco, liquids, and various other commodities, tin plate for roofing, solder, bearing metal, tin foil and tin tubes. Most of it is applied to a steel base, an operation that is best done where economic conditions are favorable. Even where the metal alone is used, as in tin tubes, it is most advantageous to produce them near the plant of the manufacturer who uses them as containers for his product. Hence, although South America is an important world source of tin, it ships the concentrated ore abroad for final treatment and imports manufactures of tin. Incidentally, the region that produces tin in South America is one of the least favorably situated for industrial development.

Sodium nitrate, of which Chile possesses a world monopoly of the natural product, is fully discussed in the chapter devoted to that country. All that need be said here is to note that the uses for it are such that there is little market for it in South America, and unless it were exported little reason for producing it would exist.

Any belief that South America is a reservoir of wealth

still lying virtually untouched must, so far as the mineral commodities are concerned, rest on the assumption that further great development of the minerals just discussed is easily possible. It certainly cannot rest on hopes for future great production of such things as gold, silver, diamonds, platinum and other minerals that have a high unit value, for while such things have intense appeal to human imagination, their quantitative importance is not great. The value of the world's yearly output of copper is ordinarily 50 per cent more than that of gold, and copper forms the basis of an enormous amount of general business, whereas gold serves for little more than an index of value. Also, as will be evident in the discussion of individual countries, South America's age of gold and of silver lies in the past, not in its future. The same is true of diamonds. Platinum, as will be indicated in due course, is an important mineral product of Colombia, but in value amounts to only about $1,500,000 yearly, nor does it seem certain that any large expansion of output can be expected.

Another most important consideration in the development and use of mineral resources is that those largely used in industry are characteristically heavy and bulky. Cheap transportation is accordingly essential to their use on any large scale and the topography of South America is not favorable to the cheap transport of material from the mineral-bearing areas to the chief centers of population. It is a large continent with numerous high plateaus and mountains. As always, the minerals are found mainly in the mountain areas. By reason in part of historical accident, the denser population for the continent as a whole is on the lowland near the sea. It seems probable that in the southern part at least this will be true permanently since these are the lands of chief food-producing value. In the north, however, as in Colombia, itself nearly one-third as large as the United States or all of France and Germany, and where the majority of the population lives on the uplands, the topography imposes all but insuperable difficulties to building railways, the cheapest known means of transporting heavy commodities over the land. It is estimated, for example, that to ship freight from Buenaventura, the chief Colombian port on the Pacific, to Bogotá would require the equivalent of elevating each ton three miles into the air. Brazil is approximately as large as the United States and borders Colombia, but it is clear that any large interchange of freight between the two must for long, if not forever, be by sea. It is possible to go by railway from the nitrate fields of Chile to the coffee lands of Brazil, but it is not possible to ship nitrate over the route if any regard whatever be paid to economy. This inland transportation difficulty, making each country so dependent on the sea, has been a considerable factor in building up the independence of each and it is extremely difficult to visualize conditions which would lead to the heavy internal interchange of freight that characterizes North America and Western Europe and which is, in turn, so large a factor in building up the heavy industries.

This hasty review of the minerals and these illustrations of their use will serve to emphasize how essential they have proved to be in the industrial development of the North Atlantic regions. No adequate substitutes are known and substitutes at best are usually unsatisfactory. They perform their service less efficiently and at higher cost. If, then, South America is to duplicate the industrial development of the North, it must possess equivalent resources or be in a position to exchange for them something the northern countries require. If the latter be the alternative the whole force of South American countries will necessarily be ranged on the side of world peace since world trade can flourish only when war is at least localized. In the succeeding chapters of this book the particular situation in each country is reviewed with a view to indicating what mineral resources are available and the conditions under which they have been or may be developed, and what social results of such development may be anticipated. The reproducible plant and animal resources of the various countries are of well-recognized importance. It is left to others to review them. Here we are concerned with the mineral resources, those which are non-reproducible and which in the aggregate are the base of the world's accumulating stock of durable goods.

CHAPTER II

THE LAND AND THE PEOPLE

A SCHOOLBOY once described South America as looking like a pork-chop, a not unapt simile since the great uplift of the Andes corresponds to the bony rib, while the contour of the northern and eastern margins somewhat resembles the outline suggested. It is a big chop, too, extending from about 13° north latitude to 54° south, with a length of 4,550 miles, and a breadth, from east to west, of 3,200 miles. Its greatest width is just south of the equator and nearly four-fifths of its land area of approximately 6,500,000 square miles lies within the tropics.

If it were possible to turn the southern hemisphere back over the northern one, like turning a glove inside out, Santiago, Chile, would not only fall some 700 miles farther from the Pole than Boston, but would also lie slightly to the east of the latter. It would really be more descriptive of the two continents to refer to them as southeastern America and northwestern America, but custom is fixed and may conveniently be followed. The eastern tip of South America is so far east of New York that Rio and Buenos Aires are more accessible to Liverpool than to our own chief port. This easterly situation of the southern continent makes little difference in its controlling physical conditions, but the fact that so large

FIG. 1. HYPSOMETRIC MAP OF SOUTH AMERICA

From Miller and Singewald's Mineral Deposits of South America McGraw-Hill Book Co., New York

a part lies about the equator does make a great deal of difference.

In spite of a general similarity in size and shape, the topography of South America is quite different from that of North America. The major river of the continent, the Amazon, flows east. In North America the great Mississippi valley opens out to the south. In South America the second great valley, that of the Paraná, opens to the south, while in North America the St. Lawrence flows to the east. Other differences are even more striking. The great mountain range of the Andes is much higher than the Rocky Mountains and lies much nearer to the Pacific, while the mountains that skirt the East coast of Brazil not only seem to rise almost out of the sea but attain an elevation of 10,000 feet as contrasted with the typical 4,000-foot elevation of the Appalachian chain, which also lies back from the coast, except in New England. Corresponding to the Laurentian peneplain of North America is the Guiana highlands, a plateau rising by precipitous escarpments to 8,000 feet, or nearly a mile above its northern analogue. This plateau is cut off on the north and west by the Orinoco valley and on the south by the great valley of the Amazon river; south of it, and west of the mountains that skirt the Brazilian coast, is another great highland 700,000 square miles in area, for which there is no analogue in North America. This highland pitches away to the west, gradually sinking from an elevation of 4,000 or 5,000 feet on the east to 1,000 feet where, nearly 2,000 miles from its eastern margin, it meets the valley of the Madeira river. Many of the mountain ranges marked on maps of this region are in reality flat-topped remnants of the plateau, which is deeply dissected by northward-flowing tributaries of the Amazon, and by rivers which are parallel to them but flow south into the Atlantic or the Paraná. A few small rivers flow to the east and break through the coast range.

South of this tableland and extending back to the Andes is a great plain, broken by a few desert ranges. It includes the pampas of Argentina and Uruguay that were until recent geological times the bed of an inland sea, and is noted for its great fertility.

A striking feature of South America is that it almost completely lacks the coastal plains so characteristic of North America. At Guayaquil in Ecuador a true coastal plain attains a maximum width of eighty miles, and there is a narrow coast plain in Peru, but from there south it is completely absent from the west coast. Argentina has a coast plain that merges with that of the Paraná river on the north and with the pampas on the west. Uruguay and Brazil have only small patches of coastal plain. The narrow coastal plain that borders the Guiana highland on the east and north is almost the only part of that region that is under cultivation. Westward from the Orinoco river the first coastal plain found is that near Maracaibo, Venezuela, which in turn is cut off on the west by a branch of the Andes beyond which lies the north Colombian coastal plain across which the Cauca and Magdalena rivers flow to the sea.

These topographical differences, when coupled with differences of latitude, make comparison between North and South America difficult, because climate, with its resultant economic effect on man, is affected by winds and ocean currents quite as much as by physical situation. The resulting complex relationship is described in detail in standard works on economic or commercial geography and need not be set forth here. Suffice it to say that the prevailing winds which blow over South America and the ocean currents which bathe its shores differ from those which affect the northern continent, and that these differences, combined with those of elevation and latitude, produce marked differences in climate. Sacramento, California, bears about the same relation to Baltimore as Lima does to Bahia. Yet the annual rainfall in Bahia is forty times that in Lima and the average annual temperature is nearly eleven degrees higher, while Baltimore has only twice as much rainfall as Sacramento and a five-degree lower annual temperature. Southern Chile is a region of heavy rainfall and cool climate, while that part of Argentina lying east of the range is a much dryer and somewhat warmer region. Northern Chile is one of the most arid regions of the world, while north of Peru this aridity is absent. The northeast coast of South America has a climate of rather high temperature and heavy rainfall that is fairly uniform from the northern tip of Colombia to Cape Race, while from that point southwest there is a slowly declining temperature and rainfall. Northern Brazil, bordering the Amazon valley, which one thinks of as typical tropical jungle, is actually a region subject to periodic droughts, and both south and north of the Amazon there are high, dry plateaus, characterized by hot days and cool nights. The region southwest of Rio de Janeiro is one of moderate rainfall and moderate temperature, gradually declining to light rainfall and cool temperatures to the south of Bahia Blanca. It will, therefore, be readily appreciated that it is impossible to describe the climate of South America in a few words with even approximate accuracy.[1]

The dominant topographical feature of the continent

[1] Anyone wishing to go into this subject more thoroughly should consult "South America: A Geography Reader," by Isaiah Bowman, "South America," by Clarence F. Jones, and "South America," by E. W. Shanahan, Methuen & Co., London, 1927.

in the high ridge of the Andes, made up largely of volcanic rocks and forming the most striking segment of what has been called "the ring of fire" around the Pacific. This enormous mountain chain is geologically recent, having been uplifted in late Tertiary or perhaps early Pleistocene time. It is not a single line of peaks. The great valley which forms the productive part of Chile, for example, lies between a coast range and the crest of the Andes proper which is here about 100 miles east of the sea. At the north of Chile the coast range disappears, and the main range broadens into a wide complex that is sometimes subdivided into as many as five belts; the western ranges, the altiplano (a piedmont plain), the high paramos of Peru and Ecuador, the Central range of Peru, and the eastern ranges. At the extreme north in Colombia the westernmost range turns westerly through the Isthmus of Panama while the easternmost swings farther and farther around until in Venezuela it is an east-west range. At its tip lies Trinidad. Some of the peaks of the Andes attain an elevation of over 22,000 feet and it has been estimated that 7 per cent of the area of South America is 10,000 feet in elevation or higher.

This high rib of the continent, characterized by igneous and metamorphic rocks, and the clastic products of their erosion, is the mineral region par excellence of South America. The major part of the metal output of the continent is associated with it and even the petroleum fields to a considerable extent are along its flanks. The nitrate deposits occur on the inner (eastern) side of the coast range of Chile. The most important coal deposits, those of Chile, are in narrow Cretaceous basins close to the Andes, as are those of Peru, Bolivia, Ecuador and Colombia. These deposits are in sedimentary rocks that flank the igneous core of the range. The copper, tin, silver, gold and other metal deposits of Peru, Bolivia, and Chile are intimately associated with the igneous rocks, however, and thus conform to the common belief of geologists that eruptive rocks associated with great mountain uplifts afford the best place to look for workable mineral deposits.

In eastern South America there are mineral deposits of importance of other types which occur in rocks of much greater geologic age. The Guiana highlands is one of those older regions, being made up of early Paleozoic or pre-Cambrian metamorphic or crystalline rocks that in places are covered by nearly horizontal sedimentary rocks of Mesozoic age. Where mineral deposits are being worked, the detailed geological relations have been studied, but for the rest of the region only general observations are available. On the whole, the mineral resources of this region are both encouraging and disappointing. Gold occurs sparingly in wide distribution throughout, and the rivers that have dissected the region have reworked the disseminated gold into surface deposits that have been locally profitable, though of no great extent. Diamond placers are also found. Only one important underground gold mine has so far been developed. The bauxite (aluminum ore) deposits are the ones of greatest present importance.

The Brazilian highland, like that of Guiana, is the remnant of much higher mountains that once occupied the region but have been cut down by erosion. Both are parts of a single region that has been cut in two by the Amazon river and the general geology is the same. But in Brazil the gold and diamond deposits have been much more productive than in the smaller northern highland. Two underground gold mines have attained marked importance in this region; one of them remains in operation. In place of bauxite deposits which characterize the Guiana highland, iron and manganese ore deposits that are of such size and extent as to rank among the important world resources of these minerals are present. They have not yet been extensively developed. Small amounts of many minerals are found, such as beryllium and zirconium. Flanking the highlands on the south are Permian sediments that contain thin coal beds which have been the object of much attention and attempted enterprise, so far without success. Petroleum has been long sought, but results have been disappointing and the geological conditions seem unfavorable.

The pampas of Argentina, the plains of Paraguay and Paraná rivers, the Amazon river plain, and that of the Orinoco are not regions in which one would expect to find mineral deposits except non-metallics, chiefly clay, sand, and gravel, with possibly coal and petroleum in the deeper sediments. The latter is the only mineral that has actually been found in important amount; it occurs in various places in Argentina, even more extensively in Venezuela and Colombia and there are other extensive areas in which the geologists consider the conditions favorable to its occurrence in quantity.

The sedimentary rocks that ordinarily contain coal deposits are not extensively present in the continent, and where they are found they seldom contain coal beds that are thick enough and of good enough quality to permit mining. Petroleum-bearing sediments are much more important, and are found from the north to the south, though the larger development has so far been in the extreme north. The metallic mineral deposits of greatest importance have been found in and along the Andes. The much older rocks of the Guiana and Brazilian highlands have yielded a variety of minerals in small amounts and contain, in Brazil and Venezuela, the most important iron-ore deposits of the continent. The iron ores of Chile, on the other hand, are of a different type. While the Andes uplift is the most important mineral region, its high average elevation and local extreme height make transportation expensive and create many operating difficulties. Its physiography makes much of it a cold arid region that is not only unattractive as a place of residence but one in which it is impossible to rely on local food supply, while its rather scanty population furnishes but a limited number of workmen.

When the white explorers reached the new world they found it, except in a few places, only sparsely settled. Its people belong to an ethnological group that was new to them, made up of many different races, in some places small warring tribes, and in others integrated groups controlling large areas. That part of North America north of the Rio Grande was chiefly inhabited by independent nomadic tribes thinly distributed over a great plain, and living on the low plane characteristic of nomads. In South America, on the other hand, the most important native population consisted of settled peoples. The Quichuas[2] of the plateaus of Ecuador and Peru, the Yuncas and Ancons of coastal Peru and the Aymaras of Bolivia were related peoples, all under Inca rule in the fifteenth century. They had none of the domesticated animals of Europe, except the dog, and none of their own, except the llama, but had developed a rather high type of civilization, though without a written language. In

[2] The spelling of many South American words is misleading to English readers. The pronunciation of this native word would, in English, be represented by Keshua; Guaqui is simply Waki. But the Spanish transliteration is so imbedded in literature and maps that it seems futile to try to change it.

the south the Araucanians were a semi-nomadic race of about the same cultural level as the North American Algonquins and Iroquois. Other tribes of the pampas were warlike hunters, as were the Arawaks of the Brazilian plateau. The people of the river valleys of the tropic region were low-grade independent tribes. Some of those of northern Brazil have attracted much popular attention because they had not advanced to the stage of using clothing.

The Inca people (and related tribes), on the other hand, were a mineral and metal-using people. They had never learned to produce and use iron, but they had cutting and scraping tools of bronze or native copper and had accumulated a stock of gold and silver that served for ornamental and religious purposes. They did not use it in exchange since it was the exclusive property of the state. Many of the supposed bronze museum objects from South America, as from other parts of the world, are really native copper, but analyses reveal that the Incas also made true bronze, apparently by first reducing the tin ore and fusing the resultant metal with copper.

These early Peruvians smelted their ores and melted metals in pottery furnaces and, not being acquainted with the use of a bellows, used copper tubes through which attendants blew to increase the intensity of the fire. Evidences of bronze production in parts of Argentina and Chile where no tin occurs have been discovered, but whether these were outposts of Peruvian civilization, or the effects on neighboring tribes of contact with the Peruvians is a question that may be left to archeologists and ethnologists to settle.

Many of the Peruvian settlements were on terraced outwash plains of the mountain range, at an average elevation of 8,000 feet, and they consequently did a good deal of work with stone. From simple rubble walls, like those of New England, all degrees of skill can be found in the construction of retaining walls, roads, canals, granaries and buildings up to the accurate shaping of joints and the clamping of blocks together with copper, mortar being unknown to them. The metallurgical operations presuppose mining operations for the production of the raw material, but since enough ore could be obtained at or close to the surface they were very simple and were carried on with only crude tools.

The first concern of the invaders was to secure as much of the precious metals as possible. Temples and storehouses were looted, graves were opened, and the existing mineral enterprises were operated on an increased scale under Spanish direction. Both surface and underground deposits of gold were worked and by 1544 the rich silver-bearing veins at Potosí in Bolivia were discovered. These were for a time the world's principal source of silver supply but were later outstripped by the yield from Peru. Ecuador and Colombia were also the scene of mining operations, but Chile, which the invaders reached in 1536, beyond conflicts with the natives and except in the Copiapó district, yielded little to them at first, and in later years copper, of which there was then no lack in Europe.

The early history of these regions has a special interest to us because of the eventual bearing it had on the present ethnic composition. In Argentina and Uruguay the Spaniards met with little or nothing in the way of gold and silver deposits, the natives were few and unfamiliar with such work, and were too warlike to be of use as agricultural laborers. In these countries they were driven out or killed and the present population is 85 to 90 per cent European in racial origin. In Brazil gold was not found for a hundred years and the Portuguese at first engaged in the production of commodities of vegetable and animal origin, enslaving the natives to perform the labor. Low in intelligence and physically ineffective, they soon proved unsatisfactory for this purpose; before the first permanent settlement was made in what is now the United States the Portuguese were bringing in black slaves from Africa to work their plantations. As a result the present population of Brazil[3] is only 35 per cent European, 20 per cent Negro or mulatto, 30 per cent mestizo, and 15 per cent native.

In Chile the Araucanians could not be driven out. For three centuries Spanish soldiers were kept on the march and it is said that to conquer and hold the country cost Spain more in men and money than all her other New World possessions combined. The method that succeeded is interesting.[4] It was by breeding them out. Four native women to each soldier was the standard ratio in the garrisons and some had as many as thirty. De Escobar had eighty-seven living children, but even at that did not claim to hold the record for his time. Chile today has a population that is 5 per cent native Indian, 30 per cent European and 65 per cent mixed blood, a vigorous and sturdy people who make excellent workmen and agricultural laborers.

In Peru, Bolivia, and Ecuador, on the other hand, we now find a population that is not more than 10 per cent European, and containing as much as 65 per cent native Indian in Ecuador and 50 per cent in Peru and Bolivia. The remainder is mestizo, with only a small Negro element in Ecuador and Peru and none in Bolivia. Along the coast of Peru there has been a small infusion of

[3] C. F. Jones: "South America," p. 90.
[4] Jones, *loc. cit.*, p. 115.

Asiatics. Climatic and other environmental conditions in those countries, which make living there somewhat trying to any but the natives, are responsible for this as well as the historical background.

In Colombia and Venezuela, where early mining was relatively unimportant (especially in the latter) we also find a small (10 per cent) European element in the population and a larger Negro or mulatto group (35 per cent in Colombia and 9 per cent in Venezuela) for the same reason as in the case of Brazil. The native Indian group is much smaller (11 per cent in Venezuela and 15 per cent in Colombia), the great majority of the population being mestizo.

Besides the effects upon the population group early mineral production in South America had two other effects worthy of mention. One was the immediate change in the character of mineral enterprise. The natives had no iron, and were correspondingly dependent on bronze. The advent of European invaders caused the substitution of iron tools and weapons for bronze, and led to a corresponding decline in copper mining, since there was no incentive to export it. There was then an ample supply of copper available in Europe and contemporary records indicate that copper from Hungary could be delivered in Spain cheaper than that from Chile. There were no native workings on iron-ore deposits and none of any importance seem to have been discovered; the needs for iron were supplied by imports from Europe.

As quicksilver was used for the amalgamation of gold and soon became extensively used in the recovery of silver from its ores, and two and one-half pounds of quicksilver were consumed in recovering a pound of silver, there was much demand for it and considerable amounts were sent out from Europe. In 1564, rich quicksilver deposits were discovered at Huancavelica in Peru. The Spanish government regarded this as so important that it at first attempted to confiscate the mine and finally took it over on payment to the discoverer of 250,000 ducats, and thereafter operated it under contracts. This was then the only government-operated mineral enterprise in South America. Other mineral deposits were allowed to be exploited by private individuals under payment of a royalty to the Crown, which at first tried to exact one-third, but successively reduced the royalty in order to stimulate discovery, until it eventually became stabilized at one-tenth. It differed according to the time and the place, the apparent general rule being "What the traffic would bear."

It will be seen, therefore, that while copper mining at first declined, gold and silver mining was greatly stimulated, and quicksilver mining was introduced as a new enterprise in Peru. Iron mining and smelting never developed into anything of more than local importance. Gold was mostly recovered by the age-old method of hand washing of gravels by slave labor, supplemented by the crushing and washing of gold-bearing quartz. Silver required more elaborate treatment and the rich deposits of the New World led to much progress in methods of treatment. While the Spaniards looted and carried away to Europe the accumulated stores of gold and silver of the natives, they later mined and reduced much larger quantities of both, especially of silver.

How much gold and silver was sent to Europe from America in the sixteenth and seventeenth centuries has been the subject of much study. C. H. Haring[5] considers

[5] C. H. Haring: "Trade and Navigation between Spain and the Indies," p. 167 (Cambridge, 1918). Anyone who wishes to go into the subject in detail should consult his bibliography, especially the three-volume "Histoire de l'expansion coloniale des peuples Europeens," by C. de Lannoy and H. vander Linden (Paris, 1907) and F. de Laiglesia's "Los caudales

that the remittances from New Spain reached their maximum at the close of the sixteenth century, when they were about 1,500,000 pesos annually. They gradually declined to 400,000 yearly in the middle of the seventeenth century, rose to 700,000 in 1672–1679 and again declined to less than 200,000 in the last decade of the seventeenth century. These were for the account of the Crown, equally comprehensive figures of shipments for private accounts are lacking. At times private shipments were confiscated and the owners granted perpetual annuities ranging from 3 to 6 per cent of the amounts taken. In the winter of 1556–1557 the private shipments thus seized amounted to two and one-half times the sum brought in for the account of the King. This, of course, discouraged private enterprise and led to attempts to evade confiscation by clandestine shipment.

The gold and silver from America that entered Europe in the sixteenth and seventeenth centuries was so great in amount as to constitute a tremendous increase in the monetary stock of the Old World. It flowed into Spain, and on out again. As Haring says (*loc. cit.*, p. 179), "Everything could be purchased with gold and silver, not only cloths and grain, but armies, heretics, and the Hegemony of Europe." Instead of making the nation rich and self-supporting, it operated to unfit it for manufacturing and commercial life. When Spain, in her naval wars, lost her fleets, her undoing was complete.

Someone has said that the chief objectives of the Spanish invasion were gold, glory and God. The first of these immediately raises the question as to whom did it belong? In the Inca empire there was no private property and

de Indias en la primera mitad del siglo XVI" (Madrid, 1904). For the effect of the shipments on Europe see "American Treasure and the Price Revolution in Spain, 1501–1650" by E. A. Hamilton, Harvard Press, 1934.

gold and silver accordingly were the property of the state. This was not unusual for those times but the Incas succeeded unusually well in enforcing the rule. The claim by the ruler to valuable minerals found in his dominions is a practice that is both old and very widespread. It particularly accorded with the Spanish concept of the ownership of precious metals. Alonzo XI of Spain had, in 1383, promulgated a decree which said, "All mines of silver and gold and lead, and of any other metal whatever, of whatsoever kind it may be, in our Royal Seignory, shall belong to us; and therefore no one shall presume to work them without our special license and command. . . ."

In the attempt to administer such a general claim to the yield of natural resources rulers everywhere found that minerals cannot be produced except by the expenditure of efforts that may fail of result, and that, human nature being what it is, men will not make this effort unless to offset their chance of loss there is a corresponding chance of personal benefit. If profits and loss flow to the state the operators become mere employees. They take no chances. Success brings them nothing and failure may cost them their employment. From Roman times down to the present rulers and statesmen have struggled with the problem of trying to retain for the state as large a proportion of the yield of natural resources as possible without restricting their production. The trend in the sixteenth to eighteenth centuries was toward making greater concessions to the discoverer. Pizarro started his adventure in Peru under a carefully drawn agreement with the Crown, and the later English and French explorers in North America had similar agreements which frequently conceded them a larger fraction of the fruits of their quest than the general text of the royal decrees allowed.

Probably as a result of this experience, the successive revisions which the mining laws of Spain underwent tended steadily toward greater liberality to the man who discovers mineral deposits. The royal ordinances of 1783 regarding mining in "New Spain" were perhaps the most advanced of the time, but Spanish mining law never came to the liberality of our own. One result is that to this day notions of equity and right as regards mineral lands differ north and south of the Rio Grande where the two fundamental systems of law meet. In this, rather than any difference in intent to deal justly, has lain the seed of many a misunderstanding.

Mining and metallurgical operations in South America in the sixteenth, seventeenth and eighteenth centuries had not only an important influence on the course of developments[6] there, but on technology and the legal aspects of mineral production. It is doubtful if they had any great influence on the separation of the South American countries from Spain. Rebellion broke out simultaneously, in 1810 in Mexico, the greatest mining region of the new world, and in Argentina where there had never been any mineral industry. These political developments were largely rebellions against colonial misgovernment and the absurd rule that made the native-born ineligible to hold public office. The colonials who had been able to establish an economy in the New World objected, it is true, to the draining away into the Old World of the

[6] It is outside our purpose to trace the development of the different regions of South America under Spanish rule, and the interested reader had best consult "The History of the New World Called America," by E. J. Payne (Oxford Press, 1892), "The Establishment of Spanish Rule in America," G. P. Putnam's Sons, N. Y., 1898, and "South America on the Eve of Emancipation," G. P. Putnam's Sons, N. Y., 1908, both by Bernard Moses.

products of their industry. In many of the regions of South America industry was based on minerals, either from the very beginning or from an early date. There were individual differences. Peru, Chile, and Bolivia were then and still remain essentially mineral-producing regions. Brazil was at first essentially agricultural, then for a time a great gold- and diamond-producing region, and has again become essentially agricultural. Argentina was long purely agricultural and has only recently come to have any mineral industry of consequence. Each of the countries felt the stirring of national consciousness and aspired to independence. The miners, proverbially individualists, still known as *particulares* in Chile, did their part in achieving it, but, it must be admitted, only their part.

Gold was such an important factor during the first century of European contact with South America that another aspect of its story deserves somewhat detailed treatment. Columbus was so impressed with the amount of gold he discovered in Santo Domingo that he wrote to Ferdinand and Isabella, March 14, 1493, "To make a short story of the profits of this voyage, I promise, with such small helps as our invincible Majesties may afford me, to furnish them all the gold they need." Herrera says that the mines there yielded Spain, to the invasion of Mexico in 1519, a total of 500,000 ducats of gold. The rich surface workings apparently were quickly exhausted and production in later years has been only small.

On his third voyage Columbus found some gold on the coast of Venezuela. Guerra and Nino found more the next year. In 1502, Columbus found gold in the region which for that reason was called Costa Rica. In 1513, Balboa reached the Pacific. According to Prescott, he had already been told by the natives of a land to the south "where they eat and drink out of golden vessels, and gold is as cheap as iron is with you." What he is supposed to have heard on the west coast confirmed this information, but though he penetrated some distance south of the Isthmus he never reached Peru. Pizarro's later discovery of it and the importance of gold in early development there is described in the chapter on that country. But a less well founded and strange legend of El Dorado, the gilded man, had marked influence on early exploration in South America.

According to F. A. Bandelier, who has made[7] a painstaking study of the many variant forms of the legend, the Indians living near Lake Guatavita, in the high tableland of Colombia, held the lake sacred. Each year they went through a ceremony of which the culminating rite was for their chief to anoint his body with gum, sprinkle it with gold dust, and proceed on a raft to the middle of the lake where he plunged in and washed off his golden covering. The people simultaneously threw gold into its water. About 1490, this tribe was conquered by the Muysca of Bogotá and the ceremony stopped, but the reports of it long persisted and seem to have been current at great distances.

Bastidas founded a settlement at Santa Marta, Colombia, in 1525, and shortly thereafter the king of Spain leased the province of Venezuela to the firm of Bartholomaus Welser & Co. of Augsburg. Its first German governor, Ambrosius Dalfinger, set out in 1529 to hunt for the gilded man and penetrated the Magdalena river valley, which belonged to the government of Santa Marta. He was finally driven out by the natives, but is reported to have obtained 70,000 pesos in gold. In 1537, Gonzalo

[7] A. F. Bandelier: "The Gilded Man," New York, 1893.

Quesada, advancing up the Magdalena, reached the high plateau, conquered the Tunja tribe and others, and obtained gold which was officially valued at 246,976 pesos, as well as some 1,800 emeralds. In 1538, he founded Bogotá and the same year there arrived in the region Belalcazar's expedition, which had come overland from Quito, and that of Federmann from Venezuela, who, supposedly lending reinforcements to his chief, had set off to follow the trail of Dalfinger in search of El Dorado. Belalcazar, who had been sent to conquer Quito, had also set off on his own initiative in the same quest. The legend had so grown in retelling that El Dorado was then supposed to go about constantly covered with gold.

We are not concerned here with the conflicting claims that arose from these three expeditions, nor with all the other explorations that seem to have had for their chief incentive the El Dorado legend. But a few of the more important need mention to indicate how persistent was its influence. Federmann's chief, von Speyer, fruitlessly ventured as far as northern Ecuador, and died before he was able to undertake a new expedition. His successor, von Hutten, penetrated the same region which Perez Quesada had meanwhile (1541) also explored, and then turned east, probably reaching the area between the Guaviare and Amazon rivers. While returning in 1545, he was killed by Carvajal and the German sway on Venezuela came to an end. The expedition of Ursua in 1560 down the Llamas, Huallaga, Ucayli, and Marañon into the Amazon from Peru seems clearly to have been based on belief in the legend. Proveda, in 1556, failed in a similar search. Diego de Cerpa, in 1569, and Pedro de Silva, in 1574, found death instead of El Dorado in the Orinoco region. Antonio de Berreo penetrated deeply into Venezuela only to be captured by the English in 1582.

The final chapter in the search was written by the English. Sir Walter Raleigh ascended the Orinoco in 1595, in search of the gilded king and his presumably gold-rich realm. Failing, he sent Walter Keymis back in 1596 and himself returned in 1617 in a final desperate attempt to recover wealth and prestige. Raleigh's published account of his "Discovery of Guiana," with its incredible accounts of one-legged men and enormous serpents, attracted much attention but was probably far from helpful in buttressing his dwindling reputation. The lake described by Raleigh remained on maps until its existence finally was disproved by Alexander von Humboldt two centuries later. Lake Guatavita itself was eventually drained, but very little gold was found in it.

The natural tendency is to dismiss all this with a smile, coupled with wonder at the gullibility involved in expending so much effort and money upon a quest with so little basis. But the history of financial investments in South America for the past two decades gives rise to the sobering reflection that the spirit which impelled the quest for the gilded man has apparently not altogether vanished. The impulse to believe in great possibilities needs the check of careful measurement and precise appraisal. In succeeding chapters an attempt is made to appraise the mineral possibilities of each of the thirteen countries of the continent, to that end.

CHAPTER III

COLOMBIA

COLOMBIA is the South American country first reached as one travels from the Canal zone along either the east or the west coast. It has the distinction of being the only one which has extensive frontage on both the Atlantic and the Pacific. It is also of a size not generally appreciated (444,100 square miles), being but one-quarter smaller than our own Alaska, a fact obscured to most readers by the peculiarities of map construction. Like Alaska, it is a region of high mountains and has been a great gold producer. But there the similarity stops, for while only sparsely inhabited it contains over 8,000,000 people as against Alaska's 80,000. Colombia is in the tropics rather than in the Far North, but the cool rainy days on the high plateaus, where most Colombians live, remind a visitor of Alaska in spring or autumn. Its situation in the tropics controls much of its life and industry. Man has gone far in learning how to protect himself from the heat, fevers, and insect pests of tropical lowlands, it is true, but it is not as yet practical to clear up whole countries. An individual mining camp or a single community can be made safe and sanitary but at a cost too huge to permit application of the methods generally. It follows that most of a tropical area is unsafe for peoples who do not have either an inherited or acquired immunity to many diseases. This rules out most white men, and it is the general consensus of opinion that continued residence in the tropics by any race, because of the necessary unremitting struggle against disease, saps the strength and particularly the initiative of its members. Despite the abundance characteristic of tropical countries it is the rule to find the people in such regions reduced to a subsistence level of living.

Fortunately, at least half of Colombia, while geographically within the tropics, is by reason of its altitude in the temperate or near-temperate zone climatically. The mountains and high plateaus rise above the low valleys and coastlands like islands out of the sea and have afforded from earliest times conditions favorable to developing and sustaining a vigorous people. It was in the high "sabañas" of Cundinamarca and Boyacá that the Chibchas lived before the white men came, and the Chibchas were a vigorous, well-integrated people with a considerable knowledge of agriculture and the arts and a government that commanded the respect of the Spaniards. Less well known than the earlier Mayas or the contemporary Aztecs and Incas, they were a people properly to be compared with them.

When the Spaniards came to the country in the early days of the sixteenth century, they established fortress towns on the coast, Santa Marta in 1525 and Cartagena in 1533, but these shortly became essentially entrepots and more and more the succeeding settlers made their way up the rivers and onto the highlands. Here in the cool temperate climate where grains and cattle flourished, they made their homes and built their cities, and here to this day are the chief centers of population and industry.

The mountain areas are high and rough. The valleys between are deep and humid. Cross lines of transport and travel have until now been all but non-existent. One result has been to confirm and even to accentuate the individualistic spirit which was alike characteristic of the native tribes and the Spanish newcomers, so that it remains true to this day that there is everywhere an intense local loyalty. The people of Antioquia feel superior to those of Cundinamarca, who in turn reciprocate the feeling. Similar relations obtain generally and after independence was won it took long years filled with much fighting for ideals to build out of the early New Granada, and later the United States of Colombia, the present strongly centralized Republica de Colombia. It is a marvel of good sense and wise statesmanship to have built up so strong a sense of nationality on the foundation of so many isolated communities.

Topographically, the country may be divided into a number of regions: (1) The Pacific coast area, with an insignificant coastal plain and a low range of densely wooded north-south mountains east of which is the Atrato-San Juan valley, forming a natural highway from the Caribbean to the Pacific; a region of dense, hot, steaming forests difficult of access and scantily inhabited. (2) Western Cordillera, the lowest and least continuous of the three great Andean ranges which branch out from the knot at Pasto near the Ecuadorean border, and extend like fingers on a hand to the north. Individual peaks in the Western Cordillera rise to 10,000 feet, but the passes are little more than half that high. (3) The Cauca valley, between the Western and Central Cordillera, opening out on the coastal plain to the north but extending for much of its distance fifteen to forty miles wide and navigable by river boat for nearly 200 miles; a region famous for its beauty and fertility. (4) The Central Cordillera, thirty to fifty miles wide and extending north some 500 miles as an unbroken rampart. The passes are few and high, being 11,000 feet or over. The peaks are numerous and rise to 19,000 feet. At the end the range dies down and disappears below the merged plains of the Cauca and Magdalena rivers, which form the beginning of the Caribbean coastal plain. (5) The Magdalena, a long narrow valley which is the principal but an imperfect route of access to the interior uplands. (6) The Eastern Cordillera, extending from Pasto north some 700 miles, swinging east into and across Venezuela and finally dying out in Trinidad. The range is from twenty-five to 120 miles wide and in it lie the more important of the high fertile sabañas in which live so large a portion of the population of the country. In all, it constitutes about one-third of the upland area of the country. The peaks rise to 13,000 to 16,700 feet and the valleys between lie at or about 8,000 feet. (7) The Llanos or plains, which slope from the foot of the Eastern Cordillera to the Orinoco and Amazon rivers. (8) The Caribbean coastal plain, which extends from the Eastern Cordillera west to the Panama border across the fronts of the central and western ranges. It is a hot, low-lying, partly swampy, tropical area. Rising out of this plain is a semi-isolated east-west range of mountains known as the Sierra Nevada de Santa Marta which runs west from the Venezuela border to the Magdalena.

The geology of Colombia has not been worked out in detail except in a few limited areas. The country needs badly a geological survey as a basis for mineral development. At present only most general statements can be made. There is a fundamental complex of crystalline rocks, consisting mainly of granites, gneisses and schists, which forms the core of the Central Cordillera and presumably underlies the rest of the country. It is commonly considered to be pre-Cambrian, but in fact may include rocks of ages up to Cretaceous, as has been pointed out by Tulio Ospina and others. The Eastern Cordillera consists of a great mass of Cretaceous sediments, shales, sandstones, limestone, with intrusive andesitic flows with a core of earlier rocks which is exposed and has been studied in Venezuela. To the west the Cretaceous beds pass under later Tertiary sediments which also lap up onto older rocks underlying the Coastal plain, reach up the main valleys, and form the major part of the Western Cordillera and the Sierra de Chocó beyond. The Tertiary was a period of intense igneous activity and uplift. It was then that the present mountain ranges were formed and its igneous rocks came up through both the crystalline complex and the Cretaceous sediments. This volcanic activity has continued to the present.

Colombia is, and has been for as far back as records carry us, an important mineral-producing country. It has been one of the world's great gold-producing regions and the chief source of its emeralds. It is now the third most important source of platinum. Deposits of copper, lead, zinc, mercury and other non-ferrous minerals are known but have been little mined, and information is not adequate to permit a sound judgment of their importance. The country has important oil fields and now ranks eighth in production. Coal is mined in a small way at a number of points and iron has been manufactured. Large bodies of rock salt are present and are mined by the Government. Building stone and clay are widely distributed and much used but cement is largely imported, since the two local mills are in the interior and transportation costs make it cheaper to supply the coast cities from abroad.

Aside from petroleum and platinum, the minerals play but a small part in the export trade of Colombia and in building up foreign credits for the country. The balance of payments for 1931 was estimated by Sr. Edm. Merchan, (corrected to accord with final figures for platinum), as follows, the figures being in millions of pesos:

Credits		Debits	
Coffee (net)	55.0	Merchandise	55.0
Petroleum (net)	4.0	Marine freights	2.1
Bananas (net)	5.0	Tourists	3.0
Hides	2.5	Insurance	1.5
Platinum	1.0	Consular expenses	0.3
Sundries	2.0	Sundries	0.5
Tourists, *etc.*	0.5	Foreign debt service	20.8
New foreign capital	21.0	Interest, *etc.*	1.5
		Balance	6.3

In tonnage, the petroleum exports amounted to 2,373,000, and there was little besides. The total amount of money invested in Colombian mining properties is estimated at $125,000,000, of which much more than half is from the United States and perhaps not more than $10,000,000 of local origin. Approximately 5,000 people were employed in the oil fields in 1931 with 25,000 in platinum and gold mining, 10,000 in coal and salt mines, and 4,000 in other mining enterprises. The total public debt amounts to about $27 per capita, and of national wealth $800.[1]

No attempt will be made here or in succeeding chapters to list the individual mines or even the mining districts. Those interested in such information will find it conveniently summarized in the work by Miller and Singewald to which reference has already been made. The Ministry of Industries at Bogotá has published in English a convenient summary of both the mineral resources

[1] Estimated by R. L. Tucker.

and the mining regulations, prepared by J. J. Jaramillo, mining engineer,[2] which may also be consulted.

It was gold which first attracted Europeans to the country. The Chibchas mined emeralds and the Caribbeans gold, and both used them freely for ornaments. The rumors of the El Dorado or gilded man were sufficiently detailed to draw to the high upland simultaneous expeditions from three different directions. These, after great hardship, met and established the power of Spain over the country.

Estimates of the total amount of gold produced to date are given in the following table.

Gold Production of Colombia[3]

	Fine oz.
1493–1600	4,115,295
1601–1700	11,252,760
1701–1800	15,110,849
1801–1900	12,317,544
1901–1925	4,703,487
Total to 1926	47,499,935

	Fine oz.
1927	160,757
1928	143,355
1929	158,732
1930	220,500
1931	286,168
1932	255,960
1933	298,277

The production now comes mainly from the Western and Central Cordilleras and from the placers along the rivers which drain them. Placer mining is responsible for a large portion of the annual output. Modern dredges have been placed on the Porce and Nechí rivers and sup-

[2] Department of Mines and Petroleum, Bogotá, March, 1933.

[3] Economic Paper 6, U. S. Bureau of Mines. Figures for 1927–1929 according to estimates of the Bank of the Republic; for 1930–1933 according to the Ministry of Industries.

plement the hydraulic mines and native workings on these and other tributaries of the Magdalena and Cauca. Farther west, on the Atrato and San Juan, are still other dredges and workings from which both gold and platinum are derived. The placers are along the flanks of the Central Cordilleras, in which numerous lode mines are also worked. In the Western Cordilleras and Sierra de Chocó, the many difficulties of prospecting leave it still unsettled whether workable lodes or merely numerous small stringers occur.

The principal gold mines are in Antioquia, Cauca, and Caldas, though Bolivar, Tolima, Santander, and Narino also are scenes of some activity. Some of the lode mines are very old, have extensive workings and have reached depths of well over 1,000 feet. According to Lindgren,[4] the deposits are in the main typical gold-quartz veins with minor amounts of various sulphides, and are related to the California type. They occur usually in granites or schists of probable pre-Cambrian age near intrusions of porphyry or monzonites of probable early Tertiary age. Veins of this type are frequently persistent and have yielded important amounts of gold in other parts of the world. At Marmato and Echandia, in Caldas, and near to Manizales veins of another type, formed probably nearer the surface in adesites and rhyolites, have been found. Such veins elsewhere are characteristically rich but do not often extend to great depth. Lindgren, an excellent authority, closes his review with the remark that "altogether Colombia must be considered as the most promising gold-bearing region of South America."

Silver is much less important and the trend through the last quarter-century has been steadily toward greater

[4] W. Lindgren: "Gold and Silver Deposits in North and South America," *Trans. A.I.M.E.*, vol. 55, p. 898, 1916.

subordination of the cheaper metal. Jaramillo estimates the total production of $60,000,000 value. Quicksilver was discovered high in the Central Cordillera at Quindio in 1786 and worked for a number of years, but the mines have been abandoned for more than a century and the grade of the ore seems to be low. Copper is known at a number of points, principally in the Central Cordillera, and was mined in a small way in Colonial days. More recently deposits at the eastern foot of the Eastern Cordillera have been explored, but without sufficient success to warrant attempting production. The transportation difficulties in the way of opening mines that require handling of materials on a tonnage basis have heretofore prevented even a systematic search for copper, lead and zinc deposits suitable for modern mining. It can only be said that deposits of all three metals are present; whether they are of commercial importance is not now known.

Platinum occurs widely in gold placers, except in Antioquia, but its value was unappreciated until recent years. Formerly it was separated from the gold at clean-ups and thrown away. One of the interesting incidents of the World War period was the tearing up and reworking for platinum of the material under the streets and sidewalks of Quibido, one of the older mining towns of the West. With recognition of the value of the metal, and especially during the period of scarcity while Russia was out of the market, systematic mining was undertaken and the output steadily rose from 10,461 ounces in 1913 to over 50,000 in 1925. For a time Colombia was the leading producer of the metal, but in 1924 Russia regained the first place, and recently Canada has also passed Colombia.

Exact figures of platinum production are not available. Estimates based on exports in recent years are as below, the figures being in ounces and for crude metals yielding perhaps 85 per cent crude platinum plus 2 to 3 per cent of other platinoids.

Year	Ounces	Year	Ounces
1922	45,000	1928	53,000
1923	48,000	1929	45,000
1924	51,000	1930	42,000
1925	38,000	1931	44,000
1926	45,000	1932	45,000
1927	46,000		

The mines are in the Intendencia de Chocó, a region of dense forests and rainfall running up to 300 inches in a year. The principal workings are along the Atrato and its tributaries, particularly the Condoto, which is the scene of operations for the South American Gold and Platinum Co., the leading producer. Three large electric-driven modern dredges are operated with power from a 2,000-horse-power hydro-electric plant on the Andagueda river nearby. A subsidiary also operates on the Andagueda river, which is a tributary of the Atrato. The British Platinum and Gold Corporation operated for a while on the Opogoda river and to the south is the Barbacoas region in which, along the Patia and Telembi rivers, platinum and gold are also found. Despite the introduction of dredges, it was estimated as recently as in 1924 that half the metal was still produced by individual miners, mainly Indians and Negroes, working with pans. Of the 45,075 ounces produced in 1932, it is estimated that 31,175 came from the dredges and 13,900 ounces from hand workings, mainly of the natives. The bulk of the platinum produced is shipped to the United States for refining. The amount of platinum available is undoubtedly large and under favorable market conditions the industry can probably be expanded to some degree.

Emerald mining has long been important in Colombia.

In Colonial times the gems were required to be shipped to Peru for cutting and were long credited to that country as the place of origin. The leading mines are at Muzo about sixty miles northwest of Bogotá in the Department of Boyacá, and have been described by J. E. Pogue,[5] who

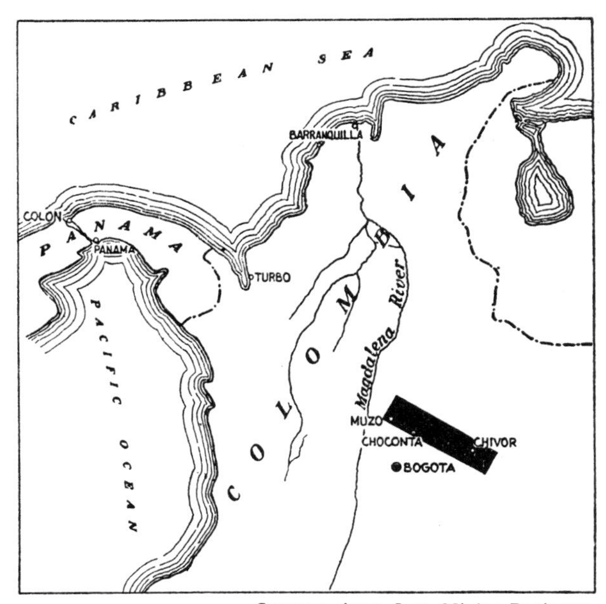

Courtesy Amer. Inst. Mining Engineers

FIG. 2. LOCATION OF EMERALD FIELDS IN COLOMBIA

found the gems to occur in limestones where small pegmatite veins have made their way into the sediments. The occurrence at Chivor, about fifty miles northeast of Bogotá, is similar, according to Rainier.[6] These and the

[5] J. E. Pogue: "The Emerald Deposits of Muzo, Colombia," *Trans. A.I.M.E.*, vol. 55, p. 910, 1916.

[6] For details see P. W. Rainier: "Chivor-Somondoco Emerald Mines of Colombia," *Trans. A.I.M.E.*, vol. 96, pp. 204–216. The geology of the deposits is described by Charles Mentzel: *Idem.*, pp. 213–216. For a

Coscues are the only ones that have as yet proved to be commercially important. The Muzo mines belong to the Government and have been worked at various times by concession or by departmental forces. Owing to the accumulation of an unsold stock of stones, little has been done at Muzo in recent years. The Chivor or Somondoco deposit is controlled by the Colombian Emerald Development Corporation, a private company under the general mining law.

The emerald mines are of considerable potential value. It is said that in certain years the Muzo mines yielded more than one million dollars' profit to the public treasury. The gem stones found are of great beauty and with proper effort a vogue for them could perhaps be created rivaling that for diamonds. This, however, requires careful and expert marketing. So far, there has been little value in the low-grade material, which naturally is much more abundant than the gem stones, and all of which requires handling. It is suggested that it be made the basis of beryllium production. The lightness and other valuable properties of that metal are slowly bringing it into use and a possible field of use for the refuse from the emerald mines may be found. How far these plans can be carried out remains to be determined, but in the emerald mines the Republic has a resource of unique value.

Coal is found at a number of points within the country. In "Coal Resources of the World," published by the International Geological Congress of 1913, Colombia is credited with a possible reserve of 27,000 million tons of bituminous coal and a like amount of sub-bituminous, thus

general account see C. K. MacFadden: "Brief Review of Emerald Mining in Colombia," *idem.*, pp. 216–223. Also a report by R. Schiebe to the Ministry of Industries.

giving it much the largest reserve in all South America. The very size and roundness of the figures mark this as but a guess, and it is a doubtful one at that. No accurate surveys have been made nor have data been collected warranting expectation of the finding of any such amount. However, the occurrence of coal is widespread, it being known at points from the Llanos of the east to the west coast. Where it has been mined, as near Bogotá, Medellin, the Cauca valley and elsewhere, it is of fair steam quality, and is found in beds of workable but not great thickness. It is broken and faulted and actual mining conditions are none too favorable. Along the Pacific railway, of which the terminus is at Buena Ventura, coal is found at a number of points and is mined in a small way. While fairly high in ash, it ranges in fixed carbon from 41 to 63 per cent and is of good steaming quality. Some of it shows possibilities of coking. The most recent report on Colombia coals is that of Emil Grosse, written in 1926.[7]

There has been no such demand in Colombia as has led to opening large mines and *per contra* there has been no such abundance of good cheap coal as to stimulate the growth of the industries which in turn create large coal demand. Railways and small local industrial plants have been supplied. Household use has been extremely limited. Neither central heating nor stoves are used even in the large cities of the upland. In Bogotá, a city of 300,000 inhabitants, a few years ago there were only, so far as could be ascertained, three coal-burning fireplaces for house heating. Since the average mean temperature even at that city on the highlands is but 60° F., this is not unnatural. Over much of the country wood or charcoal

[7] "Estudio Geologico de Terciario Carbonifero de Antioquia en la Parte Occidental de la Cordillera Central de Colombia."

is the common fuel and wood was long used by the steamboats on the Magdalena. It is now being rapidly supplanted there by fuel oil, as is generally true in the coast cities also. There is no present evidence that Colombia coal fields will ever prove to be of more than local and restricted importance, though in the western portion as a local fuel they will play their part. For that country, as generally in South America, it seems more probable that petroleum and water power will be requisitioned to do such work as requires mechanical or electrical power.

Iron ore is found at various points in Colombia but nowhere in quantities comparable to those in the great iron-producing districts of the world. As is common in pioneer countries where transportation difficulties make delivery prices extremely high, small iron works have been erected, in this case at three points, and fed with local materials. None of these so far as can be learned was of more than local and temporary importance. Of these La Pradera is best known, the mine having been operated up to 1900. None could survive the competition of imported iron and steel when adequate transportation facilities came to be provided. The only deposit for which even approximate estimates are available is the La Calera, about ten miles from Zipaquira, where William Jones and A. L. Alig are quoted by Lucien Dujardin[8] as having estimated a probable tonnage of 38,000,000 with a content of 52 to 57 per cent of iron. Presumably this was not based upon detailed drilling and sampling. While such a deposit cannot be dismissed offhand it remains true that Colombia will chiefly have to depend upon imports for its steel supply with such reworking of the material as may prove feasible when, as time passes, a stock of

[8] "Ann. de Ingeneria, Bogotá," reprinted in *Bull. Soc. National de Minera Chile*, pp. 386–391, April, 1931.

scrap accumulates in the country. Imports of iron and steel in 1929 were valued at $10,628,000, but declined to $3,128,000 in 1931.

Salt is one of the minerals which occurs in Colombia and is mined on a considerable scale by the government, particularly at Zipaquira in Cundinamarca. There is here a body of rock salt of such size as to have been estimated to contain a billion tons. It was worked as an open pit by the Chibchas but has now for many years been operated as an underground mine. The crude salt as it comes from the mine is dissolved in water in open-air tanks, and the brine, which is easily measured as to quantity and density, is run to evaporating works belonging to various concessionaires. Rock salt occurs in the Cretaceous and Tertiary formations at other points in the Eastern Cordillera but salt springs and sea water are the basis of production west of the Magdalena. The whole industry is a government monopoly and an important source of revenue.

Petroleum is at present the mineral product that contributes most to the industry and to the exports of Colombia. Mineral commodities are responsible for about 35 per cent of the total exports measured in value, and petroleum, valued at $25,169,000, corresponded to 25 per cent in 1930. In Northern South America petroleum production first became important in Trinidad. From there it spread into Venezuela, where it attained its present maximum importance, and finally on into Colombia, where in many particulars the occurrences are similar to those in Venezuela.

The general geology of the oil fields has been reviewed by Garner[9] and the origin of the oil discussed by Ander-

[9] A. H. Garner: "General Oil Geology of Colombia," *Bull. Amer. Assn. Pet. Geol.*, vol. 11, pp. 151–156, 1927.

son.[10] Hopkins[11] and Wheeler[12] have reviewed the development of the fields. Many others have written about one or more phases of them but these references will serve to guide the reader who wishes more detail.

Four regions in Colombia have attracted attention as possible sources of oil and on the western coast there is a fifth in which over a small area the rocks outcrop that farther south in Ecuador and Peru are oil-bearing. In the Catatumbo region of northeastern Colombia there is an extension of the oil fields of the Maracaibo region in Venezuela. It is a wild country, imperfectly known and inhabited in part by uncivilized Indians who have made exploration dangerous. This area contains the Barco concession; controversy with the Colombian government long hindered its development but is now happily resolved and the South American Gulf Company is drilling. Gas and small amounts of oil have been produced from a few shallow wells for some years. The area is large, the geology favorable, and it is not improbable that oil fields of major importance will be found.

The second region in which exploration has been attempted is the coastal plain stretching across the northern part of the country from the Santa Marta mountains to the Panama border. Within this area mud volcanoes, gas vents and oil seeps are widely scattered and in it a number of companies have undertaken exploration. In all, some twenty-five wells have been put down, the Repelon No. 3 of the Richmond Petroleum Co. stopping at 4,312 feet. No oil of commercial importance has yet been found despite the favorable indications.

[10] F. M. Anderson: "Original Source of Oil in Colombia," *Bull. Amer. Assn. Pet. Geol.,* vol. 10, pp. 382–404, 1926.

[11] O. B. Hopkins: see various volumes of *Pet. Dev. and Prod.,* published annually by A.I.M.E., particularly 1926, p. 725.

[12] O. C. Wheeler: *ibid.,* see especially 1933, p. 373.

Drilling conditions are very difficult and active drilling has, at least for the time being, been stopped.

The third region, and the only one which has yet attained production, is the Magdalena valley where, particularly on the east side and along the western slopes of the Eastern Cordillera from El Banco to Neiva, a distance of 500 miles, the geology is in many places favorable to the occurrence of oil. A number of companies have taken up lands in this belt and have drilled wells. So far, only the Tropical Oil Co., a subsidiary of the International Petroleum Corp. and hence of the Standard Oil of New Jersey, has developed production. O. C. Wheeler has summarized the history and status of the two fields here as below:

	Infantas	La Cira
Proved acreage	5,300	4,700
Number of wells	448	201
Average depth	1,328	3,449
Production 1932 (bbl.)	9,474,430	6,937,695
Cumulative production (bbl.)	76,708,864	42,262,655

The total production for the year 1932 was 16,416,125 barrels, and the cumulative to the end of that year 118,971,519. The average initial production in the Infantas field in 1932 was 164 barrels, and in La Cira, 603. The Infantas field has been in production fourteen years and La Cira six. The former yields oil of 27° B grade and the latter 25° B. A refinery has been built at Barranca Bermeja and a pipe line was completed to Cartagena in 1926. The larger part of the output is exported since consumption is limited in the territory which it is commercially feasible to reach from the refinery. Even so, approximately one-third of the cargo handled on the Magdalena river is that of the Tropical Oil Co., amounting to 130,683 metric tons out of a total of 444,304 in 1930 and 100,803 out of 368,221 in 1931. Since 1929, production and freight movement have decreased here as elsewhere in the world but it is clear that the region is capable of furnishing petroleum of world importance.

The fourth area in which exploration for petroleum has been undertaken is along the eastern front of the Andes and out under the great plains or llanos which stretch from there to the Orinoco and Amazon. In this area Colombia has a frontage of 650 miles between the Venezuelan and Ecuadorian borders. Here, as elsewhere at numerous points along the eastern front of the Andes, geologists find indications favorable to the occurrence of petroleum. An extensive area is considered to be within the limits of possible oil territory and exploration is under way.

As has already been stated, the Eastern Cordillera consists in large portion of Cretaceous sediments intruded by andesites and related rocks and flanked by Tertiary and later sediments. According to F. M. Anderson, who reviews the origin of the oil in a paper already cited, the Cretaceous was laid down over an ancient floor of metamorphic and crystalline rocks. Its principal divisions are the Jiron at the top, the Villeta in the middle and the Guadaloupe at the base. The upper and lower members are detrital in origin but the Villeta includes limestones which mark it as of marine origin. Despite early expectations, the Cretaceous has not here proved to be oil-bearing. It is the overlapping Tertiary, especially at the base in Eocene and possibly in the Oligocene rocks, that the petroleum is found. These are rich in organic matter and abundantly capable of serving as source rocks. In ancient embayments shales and lignite are found along the front of the ancient land from Colombia to Trinidad and such limestones as occur in the series are foraminiferal. As yet, commercial production in this area has been found only in the Tertiary.

Despite the approximately contemporaneous beginning of petroleum development in Venezuela and Colombia, production has developed much more rapidly in the former country and the output has been much larger. Oil men contend that the reason for this has lain in the different attitudes of the two governments toward the industry, the former having been unusually liberal in encouraging private development, while the latter, it is held, has been unduly suspicious and restrictive. Without going into details, assessing the causes or judging the merits of the case, it may be pointed out that in its general mining law Colombia has given the widest possible latitude to prospectors and even mere claim stakers. It is only very recently that an attempt has been made by use of the taxing powers to force the latter to work or get off the ground. As to the petroleum laws, several different codes have been enacted, each one successively more liberal than the one before. It is to be regretted that this liberalization of the law has been coincident only with the business depression and consequent period of world over-production of oil as well as the growing increase of tariff restrictions on imports. It is to be hoped and expected that with the return to normal conditions of world trade and demand such further changes as may be necessary will be introduced so as to permit that larger degree of development which the probable extent of the oil fields of Colombia would lead the world to expect. One of the considerable difficulties in the way of development is the chaotic condition of land titles over much of the country, the records having been poorly kept and many having been destroyed in the period of the civil wars. This difficulty is recognized and in the present law provision has been made for protection in the case of bona fide workers of mines.

In a later chapter the social effects of such mineral development as has taken place in the various South American countries will be discussed. It will be sufficient here to point out that in Colombia gold mining led to exploration and conquest but actual production of gold has never employed a large percentage of its citizens. The main industries of the country have long been agricultural; coffee is even now its leading export. Petroleum ranks next to coffee and its production has brought about important changes. The petroleum industry, and at Santa Marta the export banana business, has absorbed a material part of the normal increase in population. It has moved them into settled communities where they have been brought into contact with different ways of living and new wants, and the export of petroleum has built up foreign credits with which imports could be increased. Following completion of the Andean pipeline, imports of shoes moved upward. The people in the region affected had a new standard of living and the means with which to approximate it. As is customary with modern industrial enterprises, hospitals, schools, public water supplies, electric lights and many other of the conveniences and necessities of present-day life elsewhere were built in the country occupied by the oil fields or greatly expanded.

A detailed account of the effect on health and sanitation that results from the development of a petroleum deposit in tropical latitudes has recently been published[13] by Dr. A. W. Schoenlober, Medical Director for International Petroleum, Ltd. He describes the sanitary

[13] "Business Looks at Tropical Medicine," *World Petroleum*, pp. 81–84, March, 1930.

survey made of the concession granted to the Tropical Oil Co. in the Magdalena valley, Colombia. This revealed that a majority of the population of the native villages were suffering from chronic malaria, hookworm disease, and intestinal parasites. A considerable percentage had amoebic dysentery, or were carriers of the disease. Many mosquitoes capable of transmitting malaria, yellow fever, filaria, and dengue were found breeding on the property. The means taken to meet and correct these conditions can be found described in the original article. The results were that between 1920 and 1928, the time lost by employees because of sickness was reduced 60 per cent and malaria was reduced 86 per cent. During 1928, the sickness record was 35 per cent better than that among U. S. troops stationed in the Panama Canal zone. The natives who work for the company have been improved in health, efficiency, and earning power. The concession thus furnishes an object lesson as to the possibilities of public health work in the country.

With the means to satisfy them, long dormant wants of the people awoke. The same result has followed mineral development elsewhere and its general implications will be later discussed. One peculiar feature in Colombia is that the Government not only derives royalties and taxes from mining conducted by individuals or corporations but is itself in the mining business, having retained its hold upon both salt deposits and important emerald mines. In each instance it undertakes only the mining. The marketing is left to individuals.

CHAPTER IV

VENEZUELA AND TRINIDAD

VENEZUELA, with an area of nearly 400,000 square miles and a population of three and a quarter million, fronts the Caribbean Sea for approximately 800 miles. To the east is British Guiana, the westernmost of the three European colonies lying along the Atlantic between it and Brazil. On the south Venezuela is bordered by Brazil and on the south and west by Colombia. Physiographically, the country includes: (a) the mountainous areas in the northwest part and across the northern front; (b) the Maracaibo basin, lying in the northwest between two branches of the Andean mountain system; (c) the great plains of the Orinoco basin, the llanos, lying to the south and east of the mountains and forming a broad, semicircular belt extending westward from the sea to Brazil and Colombia; (d) the western portion of the highlands of Guiana, which nest in the semicircle made by the Orinoco.

The Eastern Cordillera of the Andes divides in Colombia at about Lat. 7° N., one branch continuing north along the Venezuelan boundary to the sea, and the other turning northeast into Venezuela. Here it becomes known as Sierra de Merida and forms the eastern rim of the Maracaibo basin. Beyond a low pass across the mountain axis near Barquisameto, the trend of the range gradually changes toward the east and approaches the coast. Along the coast its extension is known as the Maritime Andes. With the exception of about 100 miles between Cape Codera and Cumaná, it forms a high rampart cutting off the interior of the country from the sea all the way from Maracaibo to the delta of the Orinoco. Between the two branches of the Andes is Lake Maracaibo which lies in a major structural depression and is the center of petroleum development. The Maritime Andes, as is commonly true of other portions of the Andean mountain system, often are made up of two parallel ranges with median valleys in which, in Venezuela, lie several of the larger cities and which is now the home of the greater part of the population.

Trinidad is a large island, 1,862 square miles in area, lying just off the coast of Venezuela at the mouth of the Orinoco, which was until recent geological times part of the mainland. It is structurally the eastward continuation of the Maritime Andes of Venezuela. While it is a British colony, it may conveniently be discussed here as its principal mineral resources, asphalt and petroleum, have been intimately connected in development, as they are in occurrence and genesis, with those of the mainland.

In Trinidad, while the heights are less, the two outer ranges are clearly marked and form the northern and southern shore lines. Between is a low-lying country, itself divided into a northern and a southern basin by a middle range of low mountains or hills. The asphalt and petroleum developments are in the southern basin.

The Orinoco basin is a great low-lying plain occupying about four-fifths of Venezuela. It is open to the south so that there is water connection between its tributaries and those of the Amazon. North and west of the river itself and extending to the mountains of Colombia as well as Venezuela are the great grass-covered plains, studded with virtual islands of trees, and known as the llanos. The elevation along the rim generally does not exceed 375 to 400 feet but rises to twice that along the mountain borders. The country is so flat that in the rainy season much of it is submerged. A natural cattle range, it is now much less populous than in Colonial times, both herds and inhabitants having suffered greatly in the long period of wars that marked the winning of independence and consolidation of the country. South and east of the Orinoco the heavy forests which characterize so much of Brazil cover the country.

The highlands of Guiana, which extend into Venezuela, are a portion of an ancient land mass cut off from the main tableland of eastern Brazil by the Amazon. They will be more particularly discussed later.

The geology of Venezuela, particularly in the oil fields, has received much attention in recent years. It has been conveniently summarized by R. A. Liddle, to whose work[1] readers desiring detailed information are referred. While very old rocks occur both in the Guiana highlands to the east and the mountains on the western border, and while representatives of the Paleozoic have been recognized both in the Cordillera and in the Maritime Andes, the larger part of the country is underlain by Mesozoic, Tertiary and Pleistocene rocks. The older land masses in southeastern, western and northern Venezuela had been worn down to a low-lying plain sometime before the opening of the Cretaceous period. Over this the shallow Cretaceous sea spread rapidly, burying the folded and faulted rocks of the earlier periods. The sea deepened over most of the country and a varied series of rocks was deposited, but it was in the succeeding Tertiary era that

[1] "The Geology of Venezuela and Trinidad," 552 pages, J. P. MacGowan Pub., Fort Worth, 1928.

the coal beds were formed and the large oil reservoirs laid down. It was an era in which repeatedly the sea swept in over low-lying lands with consequent estuarine conditions favorable to burial of organic matter and accumulation of thick beds of coarse sediments. At times, extending into the Pleistocene, the land was elevated, deformed, again submerged, all of these incidents in geological history being significant as regards the formation of mineral deposits but too complicated to be followed in detail here. It is sufficient to point out that while the highlands in the southeast are old the mountains elsewhere are geologically young. Many significant structures have been developed or intensified in the most recent geological periods. The llanos and much of the rest of the country are mantled with a layer of recent deposits under which the geologically more significant rocks may be expected to be found.

Venezuela's coast was the first part of South America seen by Columbus, who traveled along it on his third voyage. In 1498, Alonzo de Ojeda and Amerigo Vespucci more thoroughly explored it. By 1550, the Captain-generalcy of Caracas was created, but no precious metals to speak of were early discovered and consequently much less attention was paid to it than to richer regions elsewhere. There were early conflicts between the Spanish, Dutch, and English and in 1617 Sir Walter Raleigh destroyed San Thome. At the same time he burned the then capital of Trinidad, but the island itself remained in Spanish possession until 1597, and was finally ceded to Great Britain in 1802. Holland succeeded in maintaining a foothold in the region and to this day Curaçao and Aruba, islands off its coast, are outposts of that country.

In 1811, Venezuela declared its independence of Spain and the great leader of its revolution, Simon Bolivar, is not only a national hero but he is the George Washington of all South America, for he followed up his success in his own country by securing the freedom of Colombia, Ecuador and Peru. Like Washington, he then wished to retire to private life, but there being little unity of purpose in the South American colonies, he found it necessary to continue in office, and was president of Colombia till his death in 1830. Venezuela was a part of Colombia until 1829. From then till 1900 the government was sometimes frankly a dictatorship, sometimes presidents were ostensibly elected after having been nominated by the real head of the government, and sometimes the power was in the hands of an oligarchy. From 1870 to 1889, Guzman Blanco controlled the country. In 1900, Cipriano Castro became President and continued in power until 1908. His presidency was marked by many conflicts with European powers and severe strain on relations with the United States. It will be remembered that President Cleveland had intervened forcibly in the dispute between Venezuela and Great Britain over the boundary between the former and British Guiana and had succeeded in securing the reference of the dispute to arbitration. Castro's attitude toward all foreign interests was such that in 1903 the English, German and Italian fleets were blockading Venezuelan ports, with the tacit consent of our government since in seizing in 1904 the property of the New York and Bermudez Co. he gave our interests the same treatment as the others. Castro was succeeded by President Juan Vicente Gómez in 1908. He has continued in power, in fact if not always in name, from then to the present. He has maintained friendly relations between Venezuela and other nations and the great development of the petroleum industry which has taken place since he came into power has provided the state with abundant revenue. It has permitted the payment in full of the troublesome public debt and furnished funds for an extensive road-building program and for other public improvements. Venezuela is almost unique in having no public debt.

As was true so generally in South America, it was gold which first attracted attention to the mineral resources of this country. Since the World War attention has been concentrated chiefly on the development of its petroleum resources, a movement greatly stimulated by the rapid growth in world demand and by the decline of the Mexican fields coincident with the period of unfriendly relations between the Mexican government and the producing companies. The discovery and development of the Venezuelan oil fields followed in natural sequence that in Trinidad and the exploitation of asphalt in both.

The mineral resource for which Trinidad is most widely known is its so-called Pitch Lake; because of its general interest the deposit is here discussed in greater detail than its present rather limited commercial importance would otherwise deserve. Near the highest part of Brea point, in the southern basin of Trinidad, and at an elevation of about 140 feet above the sea, there is a deposit consisting of about 40 per cent solid hydrocarbons, 28 per cent water, and 32 per cent fine sand and clay. This viscous mass fills the end of a valley to a depth of some 150 feet over an area of about 114 acres. Its existence was noted very early and Sir Walter Raleigh not only "tarred" the hulls of his ships with it when he visited the island in 1595, but also wrote a brief description of the "lake."

Nothing further was done about the deposit until 1830, when Admiral Cochrane sent two shiploads of the material to England. Small deposits of semisolid hydrocarbons are found in many parts of the world and have been used in a limited way for a variety of purposes from very early time. Early in the nineteenth century such material began to be used for roofing and the surfacing of floors, bridges, and sidewalks, creating a market for it that was capable of great expansion. It was so used in the United States as early as 1838 and in 1876 a pavement made from the material from the Trinidad deposit was laid in Pennsylvania Avenue, Washington. The large proportion of fine solid material in the hydrocarbon made it well adapted to paving purposes, and as a result a number of local producing units were started. In 1888 the Barber Asphalt Paving Co. negotiated a combination of these interests into a single company, the Trinidad Asphalt Co., in which it obtained a controlling interest. The terms of this agreement gave the Trinidad government a welcome revenue through an export duty of $1.20 per ton for a forty-two-year exclusive concession, and to the Barber company a monopoly of the only material then considered suitable for paving purposes. Production grew from 43,000 tons in 1887 to 96,000 tons in 1891.

Rival paving firms, thus deprived of a free source of raw material, sought other sources of supply. The New York and Bermudez Co., in 1891, obtained from the Venezuelan government the fee to some 11,000 acres of land on the western shore of the Gulf of Paria, opposite Trinidad and about 100 miles distant. Here there is a hydrocarbon deposit about 1,100 acres in extent and four or five feet thick. It contains only 3 to 5 per cent of earthy material but by that time it had been learned that it was more economical to introduce the needed amount of such material during the preparatory operations for laying the pavement, instead of paying the freight costs on its transport from Trinidad. A strong commercial rivalry developed between the two companies but was terminated in 1894 by the Barber company securing the control of the New York and Bermudez Company.

Though the New York and Bermudez Co. owned the concession and the exclusive right to ship asphalt with an export tax of 4 bolivars ($0.77) per ton, its relations with the government of Venezuela were not happy. In July, 1904, the government seized the property, on the ground that the company had not fulfilled clauses in its contract providing for digging canals, dredging river channels and exporting other commodities than asphalt. The real reason was that President Castro believed the company had backed the Matos revolt against him in 1901. Legal proceedings followed, and in 1907 the civil court of First Instances at Caracas found the company guilty and sentenced it to pay 24,178,638 bolivars (about $5,000,000) as the cost to the government of putting down the Matos rebellion. It also awarded further damages to be assessed by a commission. For the next two years the matter was under diplomatic negotiation as well as court proceedings, but was finally settled in 1909 by the payment of 300,000 bolivars to the government in satisfaction of its claims.

Although the way was now cleared for large-scale production of the Venezuela asphalt, the monopolistic control of the natural material had proved its own undoing. Rival paving interests had sought for other sources of supply. Exploratory work between the Tuxpan and Panuco rivers in Mexico led to the production of great quantities of petroleum which on refining yielded a large proportion of asphaltic residue. This proved to be as useful for paving and roofing purposes as the natural asphalt. Much petroleum of a similar quality was also discovered in California. The eventual result was that although about 4,000,000 tons of asphalt were used in the United States in 1925, only 5 per cent of it was natural asphalt, the rest being derived from petroleum. The Trinidad deposit, which had reached an output of 220,211 tons in 1913, never thereafter greatly exceeded that amount and often was below it. The production from the Venezuela deposit never much exceeded 50,000 tons yearly.

The history of these mineral enterprises is one illustration of the fact that a government which, in the belief that it possesses a natural monopoly of an essential raw material, imposes heavy duties or otherwise restricts production may end in finding that a material on which it counted for large revenues no longer produces them because of the discovery of alternative sources of supply or of substitutes. This general truth did not seem to be recognized at the time and the difficulties between the governments of the United States and of Venezuela over the asphalt deposit were a source of friction in Latin-American political relations for many years.

From 1870 on, various attempts were made to produce petroleum in Trinidad, but not until 1900 was it obtained in commercial quantity. The fine sand and clay found in the asphalt are also characteristic of the beds in which the petroleum occurs, and for a time interfered with producing the oil. In 1908, the New Trinidad Lake Asphalt, Ltd., brought in a good well, and others followed. Exports of petroleum from Trinidad amounted to 169 barrels in 1908, in 1912 to 436,805 barrels, and in 1922 to 2,444,751 barrels. By 1930, the annual output had risen to about 9,500,000 barrels. Most of the oil is refined on the island. Nearly 900 producing wells have been drilled, over one-third of them by the Trinidad Leaseholds, 175 wells by the United British Oilfields of

Trinidad, Ltd., and the rest by six other companies, mostly British. In 1930, these wells had yielded a total of 65,000,000 million barrels, or an average of 70,000 barrels per well. The average depth of drilling is 1,600 feet and the wells produce for about twenty years. Some of the petroleum rights are leased for a 10 per cent royalty, others for an annual rental of $5 to $15 per acre, or else the fee is bought for $250 to $500 per acre. Most of the wells are at no great distance (within ten miles) of Pitch Lake.

W. J. Millard has estimated[2] that the total possible reserves of petroleum in Trinidad may be as much as 1,000,000,000 barrels, which is a large amount for so small an island. Out of a total of £4,787,666 of exports in 1930, the exports of gasoline amounted to £1,189,215, kerosene £50,494, crude petroleum £222,975, and asphalt £316,397; thus these mineral commodities corresponded to more than one-third of the total exports. The other principal export products are cacao, raw sugar, and copra.

The geological history of the areas in which petroleum is found in Colombia, Venezuela and Trinidad is closely similar. The largest known fields in all three countries are related to the Eastern Cordillera and are situated along its flanks or in inter-range basins such as that in which Lake Maracaibo occurs. The significant rocks are Cretaceous and Tertiary in age and while the section differs from point to point there is a general similarity in the presence of great thickness of estuarine deposits but slightly to strongly folded and complicated with overthrusts. Millard[3] summarizes the section as developed in Trinidad as follows:

[2] *Trans. American Institute Mining and Metallurgical Engrs.*, vol. 92, p. 537, 1931.

[3] W. J. Millard: *Trans. A.I.M.E.*, vol. *Pet. Dev. and Prod.*, p. 532, 1931.

	Feet	
Upper Miocene	300–7,000	Absent on crests of folds
Middle Miocene	3,100	Oil-bearing
Lower Miocene	1,300	Oil-bearing
Oligocene	500	Oil-bearing
Eocene	1,000	Oil-bearing
Cretaceous	3,000	

The total Tertiary section in that area ranges from 5,900 to 12,600 feet in thickness and the bulk of the petroleum comes from about 3,100 feet of this. The oil sands are numerous and prolific and the possible reserves undoubtedly large.

In Venezuela, petroleum is found in both the Cretaceous and the Tertiary formations though the major production in the western and northern fields, themselves the most important in the country, is from shales, sandstones and limestones of the Upper Oligocene of the latter. The Middle Eocene yields oil in the Colon district and some is derived in Eastern Venezuela from the Upper Cretaceous but the main source in that region is considered probably to be the Miocene in horizons that may be correlated with those producing in Trinidad. The Venezuelan fields as a whole confirm the rule of the large importance of the Tertiary formations in petroleum production.

The most productive fields lie in and around the Maracaibo basin and along the northern flank of the Maritime Andes in the state of Falcón. The basin is a major topographical and structural feature of the country. It lies largely in the state of Zulia but includes parts of Mérida and Táchira, extending indeed into Colombia. To the west is the Eastern Cordillera and to the east the Venezuelan Andes. Occupying a considerable portion of the basin is Lake Maracaibo, some 120 miles long, in a north-south direction, and up to sixty miles in width. It is a shallow body of fresh or brackish water, having a maximum depth of 113 feet and being connected with the sea by a long shallow channel five to six miles wide. While there is a bar across the mouth which prevents the entrance of large ships, the fortunate position of the lake has greatly facilitated the development of the oil fields around its coast through the use of a type of tank ship adapted to its depth and of transfer stations. The oil fields, when not actually in the lake itself, are connected with its shores by short railways and pipe lines. Recently a rapidly growing network of automobile roads has come into use.

While petroleum had been recognized long before, it was not until 1884 that an organized effort was made to produce it. In that year the Compañia Petrolera de Táchira was organized by Venezuelans to work a small tract of land near Rubio in the state of Táchira. A supply developed by means of shallow-dug wells was used locally for many years. It was not until 1907 that a really important concession was granted but since then development has been rapid. The most important American and European companies have participated.

The situation of the various oil fields and prospects in Venezuela is shown on the accompanying map. The largest development[4] has taken place in and around Lake Maracaibo but it also extends for nearly fifty miles at intervals through the state of Falcón. In eastern Venezuela near the old asphalt fields petroleum is produced in the Quiriquire field, which yielded 6,756,702 barrels in 1933. At various other points, as indicated on the map, prospecting has been undertaken including the llanos south and east of the Andes. In 1928, Venezuela ranked second in petroleum production but with increasing sever-

[4] See *A.I.M.E. Pet. Dev.*, p. 247, 1932.

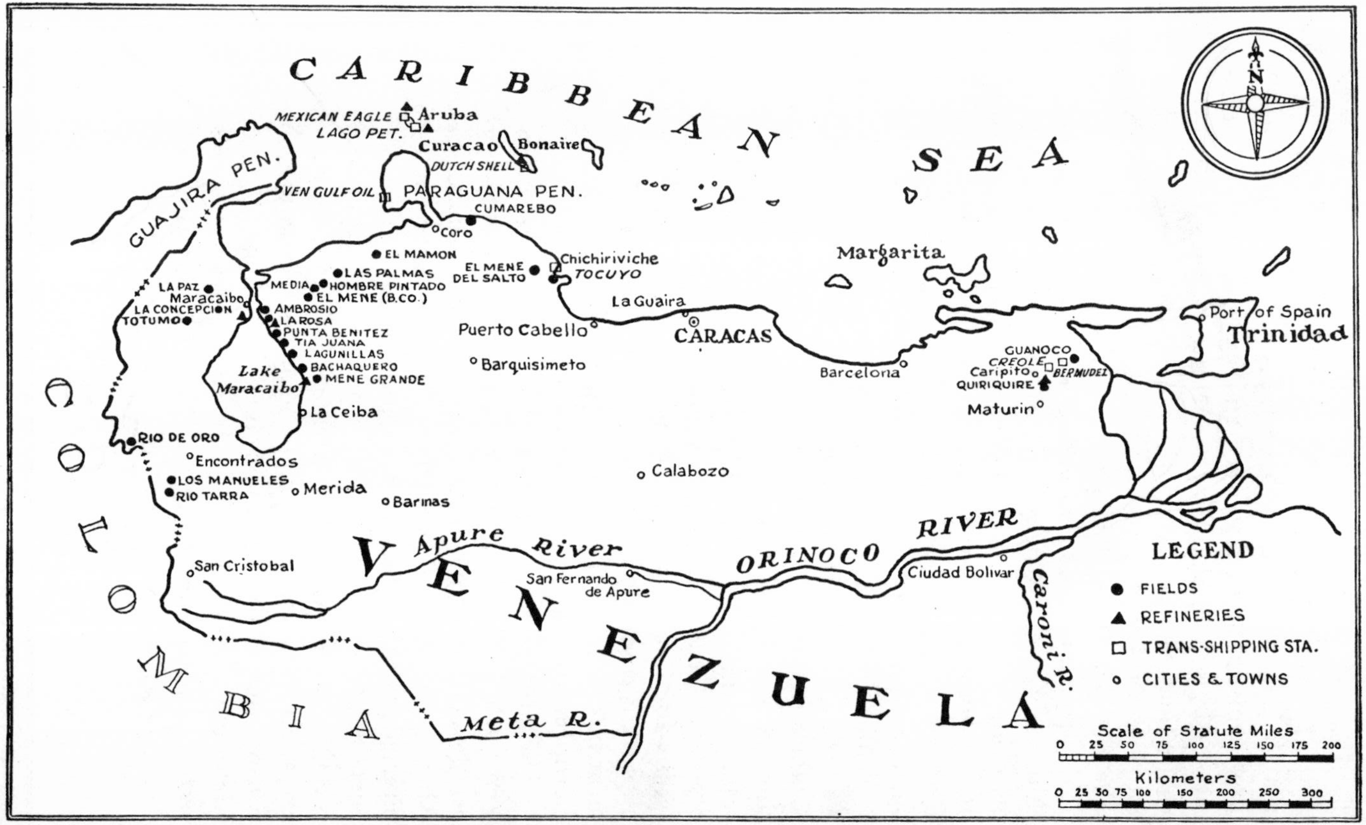

FIG. 3. PETROLEUM FIELDS OF VENEZUELA

Courtesy Amer. Inst. Min. & Met. Eng.

ity of the business depression in the western world its output fell while at the same time that of Russia rose. The latter country in 1931 resumed its historical position as the second ranking producer. The output from Venezuela in recent years has been as follows in barrels:[5]

VENEZUELAN PETROLEUM PRODUCTION

Year	Barrels	Year	Barrels
1917	226,000	1925	20,581,000
1918	367,000	1926	36,997,000
1919	259,000	1927	63,108,000
1920	526,000	1928	105,590,000
1921	1,498,000	1929	137,745,000
1922	3,394,000	1930	137,212,000
1923	3,741,000	1931	118,525,000
1924	9,147,000	1932	118,167,000
1933	120,356,000		

Substantially all of this came from the Lake Maracaibo region and of the total produced to the end of 1933, 877,339,000 barrels, 843,984,000 was exported by the leading producers to refineries abroad. The Royal Dutch-Shell company refines on Curaçao. The Lago and Mexican Eagle refine on Aruba. The remaining production formerly went to the United States for treatment but since that country levied an import tax, most of the oil has been diverted to Europe. Production which began at Mene Grande along the eastern shore of Lake Maracaibo has extended out into the lake as well as around it and also eastward along the coast. Despite the inherent difficulties of work in a hot tropical climate and the limitation on production imposed by the shallowness of the bar over which lake tankers must transport the oil, the fields have been developed with great rapidity and profit. Up to January 1, 1933,[6] 2,691 wells have been drilled in the producing fields and 188 wildcats. The ratio

[5] F. Halsmeier: *Trans. A.I.M.E.*, vol. 107, *Pet. Dev. and Prod.*, p. 445, 1934.

[6] C. C. McDermond Reports.

of dry to producing holes has been unusually low. The wells are large, initial productions being up to 15,000 barrels per day and the oils are of fair grade, ranging from 16° to 44° American Petroleum Institute. The average depth of wells is 1,315 feet in Mene de Acosta, increasing to 4,010 feet at Los Manueles. The area of possible oil territory in the country south and east of the Andes is many times that of the Maracaibo basin and the geology is sufficiently similar to warrant hope of finding considerable producing territory as economic conditions improve and the need for more petroleum becomes apparent. In petroleum Venezuela has not only a great present resource but one which offers much promise for the future.

Coal, elsewhere the world's great mineral fuel, is not important in Venezuela. In the estimates of the International Geological Congress the country is only credited with a possible 5,000,000 tons reserve each of bituminous and sub-bituminous. While these estimates are admittedly little more than guesses, they represent present knowledge. Deposits are found at various places north of the Orinoco river and are being worked in two districts. Six coal seams have been worked along the Quebrada de Araguita, which flows into the Naricual river. The mining district is connected with the port of Guanta by a railroad about twenty-five miles long. The coal is friable, yielding a large proportion of fines which are difficult to market, and the output is small, only 15,000 to 17,000 tons yearly. In quality the coal would be classed as lignite or semibituminous. In the state of Falcón coal is worked in a small way at various places and the total output is estimated at 8,000 tons per year. The coal in this region is less friable, and some of it would be classed as semianthracite. The delivered cost at the port of La Vela is even higher than coal from the

Barcelona district, which is too high to permit of competition in the coal markets of the Caribbean region. There is but little domestic market for coal, and for a long while to come, at least, it will have to compete with the fuel oil available from the large oil refineries so that prospects for the coal industry are not good.

Fortunately, the prospects for iron-ore development are much better. South of the Orinoco and near its mouth is a low range of hills, 300 to 2,000 feet high, known as the Sierra de Imataca, and forming part of the eastern highland. It is made up of a series of ferruginous quartzites, often containing nearly 50 per cent iron. These grade abruptly into itabirites similar to those found in Brazil and, as is true there, into deposits of iron ore that are of much promise. Though the deposits have been known to exist since 1900, it is only recently that they have received much study. In 1912, a shipment of 12,000 tons and in 1913 another of 57,000 tons of ore was made to the United States by the Canadian Venezuelan Ore Co. from the deposit at Manoa. Even yet the field is incompletely explored.

At Pao, which is about thirty miles south of San Felix, a small port on the Orinoco, an iron-ore deposit of workable grade occurs on top the Imataca range near its middle. Samples of ore taken from the surface and from diamond drill cores range from 67 to 71 per cent iron. W. J. Millard, in 1926, estimated that this deposit contained 12,000,000 tons of iron ore and E. F. Burchard, after further study in 1929, estimated[7] that 15,000,000 tons were reasonably assured and probably 20,000,000 tons additional were available. The average of thirty sur-

[7] E. F. Burchard: "The Pao Deposits of Iron Ore in the State of Bolivar, Venezuela," *Trans. A.I.M.E.*, vol. 96, p. 355, 1931. These deposits have also been described by Guillermo Zuloaga: T.P. 516, *A.I.M.E.*, 1934.

face samples was iron 67.64 per cent, manganese 0.25 per cent, and phosphorus 0.0355 per cent, with silica less than 0.5 per cent. Solid portions of the diamond-drill cores showed even better results. The deposit has been purchased by the Bethlehem Steel Corporation and a campaign of further drilling has proved so successful that plans are being made for opening a mine and shipping ore to the United States. The ore can be mined in open pits and easily delivered to San Felix by a railroad twenty-five miles long. It will be necessary to load it there on shallow draft boats of special type to cross the sand bar at the mouth of the river, much as petroleum is brought out from Lake Maracaibo and similarly to reload it on seagoing vessels near the mouth of the Orinoco. The ore is high in iron content and excellent in structure.

Of the six other known deposits, at Los Castillos, Piacoa, Santa Catalina, Sacuro, Manoa, and La Escondida, that at Manoa is perhaps next most important. It lies about 120 miles east of Pao; the La Escondida deposit lies farther to the east, but the others are between Pao and Manoa. As stated, some ore has been shipped from the Manoa deposit, the cargoes being reported to have averaged 66.5 per cent iron. Analyses from drill holes show 67 to 70 per cent of iron, 0.02 to 0.03 per cent manganese, 0.02 to 0.025 per cent phosphorus, 0.01 to 0.11 per cent sulphur, 0.09 to 0.50 per cent silica, and 1 to 4 per cent of alumina. It is a high grade of ore. The work done so far has not revealed whether there is a large enough mass of ore to justify the necessary capital expense for its commercial production.

Since these Venezuelan ore deposits are much nearer the coast than the Brazilian deposits and the distance from seaports to the United States is also much less, it is quite possible that these deposits will be brought into commercial development before those of Brazil. The Venezuela district is possibly an important world source of high-grade iron ore. When and if developed it will undoubtedly be on the basis of export of the ore to foreign blast furnaces, since there is neither a suitable fuel supply nor a local market for iron and steel.

Metal mining does not rank with petroleum production in Venezuela despite its much earlier start. Gold mining centers mainly in the eastern highlands. Copper and other metals are found in the Maritime Andes and are mined to some extent. There is a copper-producing district at Aroa, about fifty miles from the seaport of Tucacas. The mines have been worked intermittently since 1600 and are now operated by the South American Copper Co. Ltd., a British company. According to D. F. Haley, former assistant manager, the ore occurs as a replacement in limestone at the contact with a much crushed schist. The ore body is a great mass of solid sulphide, in places 150 feet thick and 250 feet wide. This great mass has been subjected to much faulting and on the west side a face of ore 120 feet thick containing 7 to 15 per cent copper has been cut off by a strong fault whose displacement is, as yet, not known. Large blocks of drag ore have been found all along the west side. The throw, as shown by slickensides and other evidence, is to the north and at an angle of about 18°. Development is being carried on to locate the faulted section. The ore body east of the fault was about 2,500 feet long. A 200-ton flotation plant started operation in January, 1930. Between 1924 and 1926 the company produced 1,000–1,500 tons of copper yearly, but in 1926 the output declined to 1,000 tons and to 150 in 1927. Since 1928 no official figures of copper production have been given out, but in 1931 it was unofficially reported as 600 tons. There are other known copper deposits and a long list of localities might be given, but all the evidence seems to indicate that the deposits are marginal in grade, so that after relatively small sums have been spent on exploration or working they are abandoned. Possibly they may prove eventually of importance, but the general weight of evidence is against it.

Aside from copper, there is little mineral mined in western Venezuela. Salt is a government monopoly and a considerable source of revenue. It is worked on the Araya peninsula, on the island of Coche near Barcelona, and in the states of Zulia and Falcón. The reported output is 25,000 to 30,000 tons yearly. There are small workable deposits of sulphur in the states of Táchira and Trujillo and it has been mined at Carupano. Manganese, bismuth, antimony, lead, silver, niter, and a number of other mineral substances are said to occur, but there is no reported production.

Eastern Venezuela was the scene of the first mining and is the region from which most of the gold so far produced has come. Since the plateau there is the northern end of the great ancient highland of eastern South America, it is natural that it should exhibit the same general characteristics as in Guiana and Brazil. Unfortunately, it has neither the aluminum ore of Guiana nor the manganese ore of Brazil. Iron ore, gold, and diamonds are the only ones of the characteristic minerals of this highland which appear here in any notable quantity. Even gold mining is of historical rather than of much present importance. The El Callao mine, near the border of British Guiana, was the greatest gold mine in the world in 1885, producing at the rate of $200,000 monthly. Soon thereafter it began to decline, and it has long since been closed down. It was on the Yurvari river, about 150 miles southeast of Ciudad Bolivar. Supposed to be a surface deposit when first worked by its Corsican owners about 1850, it proved instead to be a quartz vein rich in gold. Some of the material mined yielded $100 per ton. It proved practicable to work it only to a depth of 650 feet though explored to 900 feet, and about 1,000 feet in length because, although the quartz persisted beyond those limits, the gold content did not. It is now remembered not only for its period of glory, but also as being the mine where many engineers who later became prominent in the South African gold fields had their first experience. There are numerous other veins in the neighborhood. One mine is stated to have 300,000 to 400,000 tons of ore averaging $11.25 per ton.

The New Callao Gold Mining Co., Ltd., is a British company that has some fifty concessions, covering 150,000 acres, that include the old Callao property. The reported output is as follows: 2,175 ounces gold in 1921; 10,612 ounces in 1922; 2,473 ounces in 1924; no more recent figures seem available. The New Goldfields of Venezuela, Ltd., another British company, has concessions covering thirty square miles in the Roscio district, and reported a yield of 28,626 ounces from 108,244 tons in the year ended June 30, 1933. Other companies are the Botanamo Mining Co., which has 4,200 acres in the Roscio district, and reported that in 1929 it was operating at a profit of $15,000 monthly. The Bolivar Venezuela Gold Mines, Ltd., which has 2,757 acres in the same district, was taken over by the New Goldfields of Venezuela, Ltd., in March, 1930. The Cia Anon. Minero lo Increeble Nueva and the Yuruari Company are Venezuelan enterprises for which no reports are available. It appears, therefore, that while there is considerable interest in gold mining in eastern Venezuela, there has been no great output in recent years. The figures of reported output follow: 21,040 ounces in 1924; 30,540 ounces in 1925; 28,800 ounces in 1926; 46,930 ounces in 1927; 46,010 ounces in 1928, and 46,480 ounces in 1929.

The information in regard to diamond production is even less satisfactory. The stones are found on the Caroni river, and there is another small deposit on the headwaters of the Cuyani river. With the South African deposits able to yield more diamonds than world markets can absorb, it does not seem likely that diamond mining in Venezuela will be of more than local importance.

Despite the large production of petroleum and the long history of mining in Venezuela, the country remains little industrialized and is essentially one having an agricultural economy. Out of a total population of more than 3,000,000, of which statistics do not permit any accurate estimate of the number gainfully employed, probably not more than 25,000 are connected with the mines and oil wells. None the less, the minerals contribute three-quarters of the value of the exports and form the basis of support for the bolivar. Minerals have also been responsible for at least $400,000,000 investment, all of which except possibly $10,000,000 came from outside the country. Nearly 200 miles of railways, as well as numerous pipe-lines and roads have been built to furnish transportation and an extensive ocean shipping is dependent on the petroleum exports. A large sum has been paid to the government in royalties and much larger amounts have been paid out in wages and purchase of supplies within the country. It can hardly be doubted that this inflow of money has contributed notably to the political stability which has characterized the country recently.

CHAPTER V

THE GUIANAS

BRITISH GUIANA[1] (Demerara), Dutch Guiana (Surinam) and French Guiana (Cayenne) are three political divisions of South America that from the geological standpoint should not only be considered together, but should also be recognized as merely the northern end of the ancient plateau that constitutes most of eastern Brazil. Eastern Venezuela, already discussed, also includes part of this old highland. This plateau is an ancient land mass through which the Amazon river has cut a wide valley. The basement rocks which appear on its northern margin are generally the same as those on the south. Because of this basic similarity, the mineral deposits of the Guiana region are generally similar to those of the Brazilian highlands. One striking contrast exists, however, for the important iron deposits found in Brazil are absent except in Venezuela and, *per contra,* the bauxite found in Guiana is absent in the south.

When the newly discovered Western World was divided between the Spanish and Portuguese on the basis of a boundary line 370 leagues west of the Cape Verde Islands, the chief interests of the Spaniards lay well to the west of that boundary while the Portuguese were then not much interested in the western part of their zone.

[1] Also spelled Guyana and Guayana. It is apparently derived from the name of a native race.

Later they were fully occupied in trying to make effective their settlements along the Brazil coast from Bahia south. The region north of the Amazon was one that explorers of many nations coasted along; Columbus had seen it in 1498, Amerigo Vespucci and Alonzo Ojeda noted it the next year, and in 1500 Vicente de Pinzón, who discovered the Amazon, probably explored its rivers and shores. In spite of these early contacts the region has no significant history for nearly a century, when Sir Walter Raleigh ascended the Orinoco, in 1595. The Dutch first took serious interest in the region in 1598 and the French soon after. For two centuries it was sometimes in the possession of one and sometimes of another. In 1814, England and Holland finally settled their respective boundaries there, but the Venezuela boundary after long disputes was not arbitrated until in 1899. The boundary between Dutch and French Guiana was settled in 1891, while that between the latter and Brazil was only arbitrated in 1900.

Most descriptions of the Guiana region begin by dividing it into three belts: the coastal plain, an intermediate belt variously described as a plateau or as a sand and clay belt, and the mountain region. This classification seems of doubtful utility, for the coastal plains, though actually existing, are never more than fifty miles wide and sometimes less, and from the descriptions of the so-called intermediate belt it seems clear that it is in part an elevated earlier coastal plain, and in part the low marginal portions of the dissected highland now described as a mountain region. The Akarai mountains, which form not only the boundary between eastern British Guiana and Brazil but also the watershed between the Essequibo and Amazon rivers, rise only to 2,500 feet but there are various peaks that rise to over 7,000 and

Mt. Roraima (8,635 feet) rising at the point where Venezuela, Brazil and British Guiana meet with a sheer cliff face 1,500 feet high, has long been an object of interest. It would perhaps convey a more accurate picture of the region as a whole to describe it as an old plateau now so dissected that only short ranges and isolated peaks represent its original general level. Being in a region of abundant rainfall, it is drained by numerous rivers that, beginning with the Oyapok, at the eastern margin of French Guiana, all flow north until one passes beyond the Essequibo, which drains the whole central part of British Guiana. The Essequibo is joined near its mouth by the Mazaruni and Cuyuní, which flow east from the highlands lying north of the Mt. Roraima region; these rivers parallel the Orinoco in their direction of flow. The general slope is northward, except in the northwest corner of British Guiana, where the slope is eastward. The region lies between 1° and 8° N. latitude and 52° and 60° W. longitude.

Being so near the equator and with a rainfall on the coast of eighty inches annually, the region is covered with abundant tropical vegetation and about the only access into the interior is by the rivers. In flood time these so penetrate the forest that there is little shore on which to land. In the two dry seasons (September to December) (February to April) the rivers shrink so rapidly that travel is difficult, and in places is all but impossible because of rapids where they cross rocky ridges. It is natural, therefore, that most of the people of the region live on or near the seacoast, even including native tribes. Some basic data in the table below are given only in round figures because of the uncertainty attaching to them. The number of natives living in the forests is generally unknown.

	Area sq. mi.	Population
British Guiana........	90,000	350,000
Dutch Guiana.........	55,000	150,000
French Guiana........	35,000	50,000

The estimate of population for French Guiana does not include aborigines, but that area is the least thriving as well as the smallest of the three, having been originally a penal colony. In all of these countries there is a remarkable mixture of races, notably in Surinam, where in addition to some 2,000 whites there are coolies from Japan, China, Java, and India as well as many Negroes. In British Guiana two-thirds of the population are either Negroes or East Indians, the Portuguese, about 10,000 in number, are twice as many as the other Europeans, and there are probably less than 10,000 aborigines.

Gold and diamonds have been the most sought-after of mineral substances in this region. There is an abundant literature on the subject of gold mining. In fact, it is doubtful if any other region has so high a ratio of literature to actual gold produced. In British Guiana no gold of any importance was found before 1863 and as late as 1885 the annual output was less than 1,000 ounces. It then steadily increased to a high-water mark, 138,527 ounces, in 1919. In spite of its much smaller size the output in French Guiana has recently been more important. Some statistics follow:

	1900	1910	1918	1927	1928	1929	1930	1931
British Guiana	114,000	55,000	24,546	6,720	6,070	7,270	6,933	
Dutch Guiana	27,000	37,000	16,000	7,710	5,500	3,570	4,770	4,600
French Guiana	68,400	123,168	80,477	43,500	45,500	49,000		

One reason for variability of these figures is the varying success, from year to year, of the few large undertakings that contribute most of it. In British Guiana about half the present output is from the dredging operations of the Minnehaha Development Co., Ltd., which reported an output of 3,851 ounces in 1929, and a total to that date of 68,800 ounces. The Guiana Gold Co., which operated a dredge on the Essequibo river and claimed an output of 91,500 ounces between 1906 and 1927, has not reported any since the latter date. The rest comes from small native operations. In Dutch Guiana all the present output is from native workings; curiously enough, the principal foreign enterprise is a French company, Cie des Mines d'Or de la Guyane Hollandaise, that has been taken over by a British company, Neotropical Concessions, Ltd. In French Guiana all the output appears to come from the operations of a single company, Société Nouvelle de Saint-Elie, at Adieu-Vat on the Sinnamary river, where underground work has been carried on at various times since 1878.

During the period between 1890 and 1900 there was much interest in the possibilities of underground gold mining in British Guiana and many enterprises were started, mostly in the northwestern part, but all of them have long since ceased operation. Since a prospecting license can be obtained for $5 and gold-bearing land leased for twenty cents per acre per year (ten cents per acre for dredging land) the industry does not languish because of official hindrances. In spite of the usual conclusion in the reports of explorers that there are large unexplored areas and great possibilities, the fact should be squarely faced that the region has been one in which gold-seeking has persisted over a long period and no large or very profitable deposit has yet been found. The total production in British Guiana to date has been estimated at $50,000,000, which seems conservative, as the output since 1900 has been over half that sum. The output in French Guiana since 1900 has also amounted to about $56,000,000, but that from Surinam has been only one-quarter as much. The totals to date may be estimated as: Surinam, $25,000,000; British Guiana, $50,000,000, and French Guiana, $100,000,000. It is entirely possible that further exploration may discover a large and important underground mine, though the chances seem much against it. The placer-mining possibilities have been sufficiently explored so there seems no probability of any large new discovery. The railroad that was built 100 miles inland from Paramaribo about twenty-five years ago was mainly based on a hoped-for development of gold production in Dutch Guiana that failed to materialize.

As in the Brazilian region to the south, placer-diamond deposits are found in the Guianas, and it is probably safe to estimate that the total value of their output in British Guiana has been at least equal to that of gold. The official figures for 1928 were 132,482 carats, valued at £523,007. The total production for 1901–1930 is stated to have been 1,766,227 carats, valued at £7,256,776, and for 1930 110,042 carats, valued at £323,826. In view of the decision of the immensely larger and more productive British-controlled diamond mines in Africa, made in 1932, to cease all production for two years, there seems little likelihood that the industry will much expand in the near future. Owing to the nature of the deposits they can be worked only on a rather primitive scale with native labor. On the other hand, the local importance of diamond production in recent years must not be overlooked, for the value of the diamond output in 1928 was over one-sixth of the total exports in that year.

The raw material from which aluminum is made, bauxite, is the principal mineral (from the tonnage standpoint) produced in British and Dutch Guiana; the deposits do not seem to extend into French Guiana. The deposits are laterites, or ores that have been formed by the decomposition and weathering of the original rock at the place where they are now found. In size they vary from small lenses to deposits several hundred yards wide and several miles long. Generally they form low hills, and this feature facilitates their discovery. The workable deposits are commonly found in the intermediate zone between the coastal plain and the foothills, but in the foothill belt large low-grade deposits also occur capping some of the plateaus. These latter are of no present commercial importance since they are not only low in aluminum content and too high in iron and titanium, but also unfavorably situated as regards transportation. The high-grade deposits now being worked are convenient to river transportation.

Only one company is yet operating in British Guiana, the Demerara Bauxite Co., Ltd., organized in 1916, now a British Guiana subsidiary of Aluminum, Ltd., a Canadian corporation affiliated with the Aluminum Co. of America. The mine is at Three Friends, and a ten-mile narrow-gauge railroad connects it with the washing and shipping plant at Mackenzie, on the Demerara river. Ocean-going steamers can ascend to this point. The bauxite, which averages nineteen inches thick, dips to the east, and the overlying material, which reaches a thickness of sixty feet in places, is removed by stripping, mainly with steam-shovels. The washed ore is calcined in kilns to drive off the combined moisture and thus save freight charges. The coal for calcining is brought from the United States. Dock facilities permit 2,500 tons of bauxite to be loaded on a steamer in eight hours. About 800 men are employed on all the operations. The recovery is reported at 70 per cent. To the end of 1930, 1,280,000 long tons of bauxite had been shipped and the exports that year were 119,616 tons, valued at £124,600. In 1929, the British and Colonial Bauxite Co., Ltd., obtained a lease on another bauxite deposit on the Demerara river, near Christianburg. No production has been reported.

Of the bauxite deposits being worked in Dutch Guiana, one is at Moengo, on the Cottica river. The operating company there is the Surinaamische Maatschappij, a Dutch subsidiary of the Aluminum Co. of America. The bauxite here averages eleven feet thick with up to ten feet of overlying material that must be carefully cleaned off by hand to avoid its becoming mixed with the bauxite and lowering its grade. The ore is then blasted, loaded into cars by steam-shovels, and taken to a washing and calcining plant. Moengo is about 100 miles from the sea, but the river is deep enough for vessels of seventeen feet draught. The average vessel load in 1928 was 2,600 tons, seventy-three vessels having been loaded in that year. In 1930, 253,866 tons was exported, but only 179,107 tons in 1931. The other bauxite producer in Dutch Guiana is the Kalbfleisch Corporation, which exported 11,489 tons in 1931 as compared with 9,149 tons in 1930.

Aluminum being a light metal, it is inevitable that any ore will not contain a high percentage by weight. Even after calcining, the material shipped from the Guianas contains only a little more than 30 per cent aluminum, but it is as good as can be produced elsewhere. The Guiana material has the advantage of being low in silica, that from British Guiana averaging 2½ per cent and from Dutch Guiana only 2 per cent. This is about one-quarter of the silica content of the material produced in the United States. Bauxite in its natural state contains water which is combined with it, not simply absorbed. It is driven off by heating to a sufficiently high temperature. The weight thus lost by calcining is rather high in the Guiana material, but the saving in freight charges is more than the cost of the operation. Conversely, it is not commercially practicable to ship the raw material.

Guiana, in 1929, produced nearly one-fifth of the world's supply of bauxite, taking second rank in the world's producing regions. Its production is therefore not only important locally but is an important factor in the world situation. The known resources there indicate that the region can continue to produce at its present rate for thirty or forty years, and the additional exploration that will doubtless take place in the interim quite possibly may add much to its reserves. There seems no possibility that a local aluminum-producing industry can be developed, for the reduction of aluminum requires enormous amounts of electric energy and can be done economically only where this electricity can be produced at a low cost per kilowatt hour. In a region where the prevailing temperature is high, much attention must be paid to health and sanitation, and where the labor available is low-grade, operating conditions are also not attractive. Since water transportation, which is relatively low-cost, is available, it will doubtless remain more economic to ship raw material, even though containing only 25 to 30 per cent of the metal, to regions where reduction plants of high efficiency are available

Iron ore occurs in British Guiana and the deposits are reported to be extensive. They are lateritic ores of the same general character as those found in Cuba. Ores of this type are not attractive to ironmakers unless ore of better quality is not available, and while it is possible that they may in the future prove of value, they are not of immediate importance. A quicksilver deposit occurs in the Maroni district, Dutch Guiana, and some exploratory work has been done, but no production is reported.

Traces of petroleum and natural gas have been reported from wells bored near the coast of Dutch Guiana, but the indications that they exist in commercially valuable quantities is slight. In French Guiana the occurrence of copper, lead, tin, iron and manganese minerals has been reported; there is also a report of a deposit of coal near the Maroni river. The evidence so far available does not indicate that these deposits are of much present importance.

It is evident from the foregoing that topographical, geological, and climatic conditions make the coast and the areas bordering on navigable rivers the only parts of the Guianas that are now commercially exploitable. Gold and diamonds, because of their high value per unit of weight, are to some degree an exception, but in their case as well the difficulty of carrying on exploratory work in the interior has hampered development. Experience to date indicates that gold and diamonds, while of considerable local importance, are not likely to give rise to large or permanent enterprises.

Possibly their chief social value is the incentive they supply for exploration of the less accessible portions of the interior regions. The bauxite deposits are large and valuable enough to be the basis of important enterprises that have an assured life and may considerably expand. The racial make-up of the population of the region, combined with natural conditions, indicates little probability of rapid advance along industrial lines.

CHAPTER VI

BRAZIL

BRAZIL is almost half of South America in area; 3,286,170 out of an estimated total of 7,250,000 square miles. In population it is more than half, or 41,478,000 out of an estimated total of somewhere between 75 and 80 millions. It is, indeed, from the standpoint of area, one of the five great countries of the world and the only one of them that lies almost wholly in the tropics. Fifteen times as large as Germany, it has only a little more than half its population. Only 3½ per cent[1] of its area is under cultivation. Half, or perhaps even more, of the people are illiterate Indians, Negroes, or half-breeds who live on so low an economic scale as to scarcely count for anything from the industrial standpoint. In 1808, the population of Brazil was estimated at 3,000,000; one-third Negro slaves, one-third free Negroes and mestizos, and one-quarter Europeans. By 1900, its population had reached 17 1/3 million and at present is estimated at one-third European, one-third mestizo, and one-third native Indian, Negro, and Indian-Negro half-breeds. Brazil, until recently, benefited but little from European immigration; more European immigrants poured into the United States in the three years 1905–1906–1907 than had gone to Brazil in the pre-

[1] C. F. Jones: "South America," Henry Holt & Co., p. 410, New York, 1930.

ceding century, but between 1920 and 1925 nearly 4 million immigrants came to Brazil. The population increased more in thirty years after 1900 than it had in five centuries preceding. A few words about its history will conduce to an understanding of the economic status of its people, and consequently of their need for and use of minerals.

In 1500, while en route to the Cape of Good Hope, the expedition of Pedro Alvarez Cabral was driven by contrary winds across the Atlantic and landed on the coast of Brazil in what is now southern Bahia. A few months earlier Columbus's companion, Pinzon, had come as far south as Pernambuco. Cabral claimed the country for Portugal and the same year an expedition under the Italian explorer, Amerigo Vespucci, was sent out to report on the region, since Portugal had been awarded, in 1494, all new lands less than 370 leagues west of the Cape Verde islands. Vespucci systematically explored the coast of Brazil for 2,000 miles and reported that the only natural resources of immediate interest were immense quantities of trees that would yield the red dye known in Portugal as "brazil" (from a word meaning a burning coal). Cabral had called the new land Vera Cruz but from being known as the region whence the brazil dye-wood came it was eventually designated Brazil. The trees (species of Caesalpina) were not the same as the source of the original brazil dye, being native to America, so it is somewhat difficult to say whether the country derived its name from the dye-wood, or *vice versa*. Incidentally, Vespucci was convinced that the region he had explored was too extensive and lay too far south to be a part of Cathay and thought it must be a "new world." As a result, Ringmann, in 1507, proposed that it be called Amerigo, and Waldseemüller's map of

1508 used the name of America for the new world, although Columbus had really discovered it.

As the Portuguese were occupied with their African discoveries they paid little attention to the new region at first, although a settlement was made at Bahia in 1502. But in 1526, the King of Spain sent Sebastian Cabot to locate the meridian that was Spain's eastern boundary in the newly discovered region. When he put in at the Plata river he heard of a rich people who lived near its headwaters, so he landed and spent three years exploring along the Uruguay and Paraná rivers, penetrating as far as what is now Asunción. This seemed to be Spanish aggression in Portuguese lands, while the French had also been poaching on the coast of Brazil in the quest of dye-wood. Portugal therefore resolved to systematically occupy the claimed region. Souza established a permanent settlement at São Vincente in 1532, and the entire coast was divided into fifty-league sections, with rights extending indefinitely inland, and granted to adventurers. Only six proprietors established colonies and in 1549 a governor-general was appointed for the whole region. Sugar cane had been introduced as early as 1525, and became the principal export commodity. For labor on the plantations reliance was first upon the natives, belonging to four great groups, who ranged in character from such people as the Botucudos, so primitive as scarcely to be said to have advanced to even the earliest stages of human civilization, to the Arawaks of the hinterland plateau, who might perhaps be compared to our Apaches. They were poor material from which to develop agricultural laborers, and the supply was, in many regions, scanty. Consequently, expeditions into the interior had for their principal objective the rounding up of natives and bringing them down to the coast to work on the plantations as slaves. In 1554, São Paulo was founded, and became the center from which *bandeirantes* set out in all directions, ostensibly to Christianize the natives, but really to bring them in. Rio de Janeiro, settled by the French in 1558 and occupied by the Portuguese in 1567, was an important slave market. In 1560, slave-hunters discovered gold at Jaragua, in São Paulo, but not in important amount.

About 1600, the importation of black slaves from Africa began and continued for over two and one-half centuries; between 1825 and 1850, over 1,250,000 are said to have been brought in. It is perhaps a little misleading to refer to the Portuguese who came to the new region as settlers. They came to make profits rather than a new home. Leaving their women at home, they cohabited with the natives and later with the African slaves. A recent publication of the Companhia Energia Electrica da Bahia says "The influence of these black peoples from the primitive forest life of Africa upon the colonial civilization of Bahia is simply incalculable. It is difficult today to determine which is of Portuguese and which of Negro origin. The simple processes of agriculture, the manner of preparing food with palm oil and herbs and spices, the dances, the music, the folk-lore, are all evidences of the insidious culture of these highly complex, if primitive, people." The first Negroes came from Angola and Mozambique, but eventually many different races were introduced.

At a date variously given as 1693, 1695, and 1699, gold in important amounts was discovered in Minas Geraes, and the demand for labor became more acute. A quarter-century later, diamonds were discovered. While these discoveries greatly stimulated immigration from Portugal, the conditions were basically different from our own California gold rush. The treasure seekers of 1849 expected to, and did, produce the gold by their own personal efforts. No low-grade labor was ever brought into the region, and the Chinese and others who came in of their own initiative were relatively few in number. In Brazil, on the contrary, the work of gold and diamond digging and washing was typically done by low-grade laborers, slaves and half-breeds.

Minas Geraes, with an area of 188,000 square miles and a population (1916) of 5,000,000, naturally invites contrast with California, which has an area of 158,000 square miles and a population (1930) of 5,677,250, since both owe their stimulus to effective settlement to the discovery of gold, though the discovery in Minas was more than a century earlier. The natural conditions in Minas must first be described.

Brazil may be easiest visualized as an area 250,000 square miles, larger than our forty-eight states combined and of the shape of a wide piece of pie. Its seacoast corresponds to the rim of crust quite exactly, for what is ordinarily described as the coast range of mountains is only the high crenellated rim of the plateau behind it. To the northwest and southwest the surface declines and the Amazon and Paraná rivers correspond to the cut edges. These rivers are, of course, low-lying and the northern boundary of Brazil lies a considerable distance north of the Amazon, cutting across the southern margin of the smaller Guiana highland to the north, which was once an integral part of the Brazilian highland before the Amazon sawed its way through. If the extensive, but largely undeveloped, region north of the Amazon is left out of the picture, the triangular contour would be exact, one tip touching Peru at the headwaters of the Amazon, and at no great elevation, another tip at Cape Race (São

Roque) and the third at the junction with Uruguay and the Atlantic Ocean, the directions being from Peru east to Cape Race, thence southwest to Uruguay and from there northwest to the place of beginning. The first and last margins are low-lying, while from the Atlantic coast the rim rises steeply, ending in peaks that are the highest remnants of an originally much higher plateau. Occasional peaks reach as high as 10,000 feet, but most of the area is below 5,000 feet and it slopes away to the northwest and southwest.

This region is one of the oldest land areas of the world and the basal part of it is made up of ancient crystalline and metamorphic rocks. Some of the latter may belong to the early Paleozoic era, but the evidence is uncertain, and in general the structures in the pre-Paleozoic rocks are so complex that they have been deciphered only in local areas. At intervals the region has been lowered and later deposits laid down on it, but it has been a land area since Tertiary time. In contrast with our Rocky Mountain region, where the ranges have been formed by folding and faulting, the mountains in the region typically represent the parts most resistant to erosion and some of the "mountain ranges" which appear on maps are merely the dissected edges of escarpments of the plateau.

Typical of the region is the São Francisco river, which rises to the northwest of Rio de Janeiro, not very far back from the coast, flows north, in a valley that averages about 1,000 feet below the general elevation of the region, until it is some distance north of Bahia, and then turns east to the Atlantic. Between its headwaters and its mouth are numerous small rivers flowing east to the ocean.

The rivers to the west of the São Francisco all flow

north, either directly into the ocean along the coast as it trends northwest from Cape Race, or else into the Amazon. South and west of the headwaters of the São Francisco, the rivers flow either west or southwest into the Paraná, or the Uruguay. Minas Geraes, therefore, is the highest large area of the highland with a general elevation of about 3,000 feet, but there are peaks in Bahia that are twice as high. About a quarter of the cultivated land of Brazil is in Minas Geraes, while nearly a third is in the adjoining state of São Paulo; these two together with Rio Grande do Sul (which immediately adjoins Uruguay) and Bahia, which adjoins Minas on the northeast, contain three-quarters of the cultivated land of Brazil.

It can be readily appreciated that this highland is a region difficult to penetrate by railways. Going down from São Paulo to the port of Santos, less than fifty miles to the east, one gets a vivid impression of this on reaching the edge of the escarpment and viewing the port only a few miles away but far below. Here an inclined railway is used, but the other lines inland from the coast, mostly radiating from Rio de Janeiro, climb by steep grades and many curves back into the hinterland. A standard-gauge road now extends from Rio to Bello Horizonte, the state capital of Minas, but is of relatively recent construction, and it readily can be appreciated how difficult of access this region was in an earlier day. The rivers not only flow great distances before reaching the coast, but have their courses so interrupted with rapids and falls as to not be useful as a means of ingress.

The eastern margin of the valley of the river São Francisco and its tributaries in Minas is a ridge of high plateau, called the Serra de Espinhaço, running from south to north, and at Diamantina splitting into two branches with the valley of the headwaters of the river Jequitinhonha between them. East of Bello Horizonte the headwaters of the river Doce, which flows east to the Atlantic, drain the eastern side of this ridge. On both the eastern and western flanks of the ridge were found alluvial gold deposits, and it was the working of these that began at the end of the seventeenth century. At first the deposits adjacent to the streams were worked and then long ditches were constructed to convey water so the higher-lying gravels could be worked. The traveler through the region can observe turned-over surfaces everywhere, mute evidence of the extent of the early workings.

No very reliable accounts of this early work now exist, if indeed they were ever available. No production records were kept and the amounts can be estimated only by multiplying by five the sums collected as the "royal fifth" of the output. Since these sums include the amount of "bootlegged" gold that was detected and confiscated, but do not include that which escaped detection, they are only a rough index of the production. Estimates reached in this way indicate an annual output of nearly $1,000,000 between 1700 and 1715, rising to about $1,500,000 annually between 1735 and 1750, and attaining $1,750,000 yearly between 1750 and 1775. A rough guess is that $250,000,000 worth of gold was produced in the seventy-five years following 1700, which, though only a modest amount as compared to the $1,800,000,000 California yielded between 1849 and 1926, was equally as important as a world factor at the time. During the seventeenth century, Colombia had been the leading gold-producing country and is estimated[2] to have yielded $220,000,000 worth of gold between 1600 and 1700, or 40 per cent of the world output for the century. Between 1700 and 1800, Brazil is estimated to have yielded $540,000,000, or 44 per cent of the world total for that century, taking first place from Colombia, though the two of them produced nearly 70 per cent of the world's gold in the eighteenth century.

About 1720, diamonds were discovered near Diamantina by gold-washers who were steadily pushing northward from the original gold discoveries, near Ouro Preto.[3] The accepted story is that they were at first not recognized as having any value and the few that were saved were presented to the governor of Villa do Principe, who used them as counters in card games. Eventually some made their way to Holland, where their value was recognized. Their discovery must have been as early as the date given, for the commander of a Dutch ship records that some of his sailors deserted the vessel in Rio Janeiro in 1721 in order to go to the newly discovered diamond mines.[4] The remainder of the eighteenth century was, therefore, a time of both gold and diamond production in Minas Geraes, but there is not much available in the way of authentic accounts of it.

In 1809, John Mawe, an Englishman who was the author of several books on mineralogy and geology, visited Brazil and secured from the Prince Regent official permission to visit the diamond and gold workings, which, he says,[5] was "a favor had never as yet been granted to

[2] Summarized data of gold production, U. S. Bur. of Mines, Economic Paper No. 6, 1929.

[3] Ouro Preto was long known as Villa Rica and Diamantina as Tejuco.

[4] "Kerr's Voyages," vol. 11, p. 80, 1814.

[5] John Mawe: "Travels in the Interior of Brazil," 2nd Ed., p. 195, London, 1821.

a foreigner, nor had any Portuguese been permitted to visit the vicinity where the works are situated except on business relative to them and even then on restrictions which rendered it impossible to acquire the means of giving an adequate description of them to the public." He set out, on muleback, from Rio de Janeiro August 17, 1809, and reached Ouro Preto (200 miles distant in a straight line) twelve days later. The only means of transport in the region was by mule trains and in the course of his narrative he mentions that producers of sugar at Minas Novas were accustomed to give half to two-thirds of the sale price of sugar at Rio de Janeiro to transporters, who commonly took two to three months to make the 400-mile journey out and as long to return. He also mentions registry offices along the route at which goods being taken in had to pay a duty, which he says was about 100 per cent of the value in the case of salt, iron and lead, while textiles paid only about 10 per cent.

Ouro Preto was then, and long had been, the capital of the Province. Mawe describes it as containing 2,000 "habitations". . . . "but by no means so well peopled as when the mines were rich." (He means when the workings in the immediate vicinity were at the height of their production, 1730–1750.) He says that many of the houses were untenanted, rents were continually lowering and dwellings were offered for sale at half what it cost to build them a few years before. In his description of the places he visited through the region many are portrayed as partly deserted, and his general picture is one of decline, which seems somewhat inconsistent with his estimate[6] that the total gold production of Minas Geraes in 1809 was $7,000,000, for, while the output in individual years may have been larger, the annual average

[6] *Loc. cit.*, p. 384.

for the eighteenth century was certainly less than that amount. He does not mention seeing the official records of the "royal fifth" and probably accepted verbal statements that may have been considerably exaggerated both as to former glory and present importance.

The picture which he gives for both gold and diamond production is of surface washings in which the work was performed by slaves with only the simplest of mechanical aids, and he made frequent suggestions for improvements in this and other kinds of work which he saw going on. He also tells of showing various family groups how to make butter and cheese, and even to brew beer, but was shrewd enough to doubt whether his suggestions bore any permanent fruit. "This aversion to improvement I have often observed in the inhabitants of Brazil; when, for instance, I questioned a brick-maker, a sugar-maker, a soap-boiler, or even a miner, as to his reasons for conducting his concerns in such an imperfect manner, I have been almost invariably referred to a Negro for answers to my interrogations."[7]

In the appendix to Mawe's volume is a section headed "View on the State of Society Among the Middling Classes, Employed in Mining and Agriculture." He pictures a man who owns fifty or sixty Negro slaves worth $6,000 to $7,500 and equipment for the simple working of gold deposits as living in a house "consisting of a few apartments built up to each other without regularity; the walls wickerwork filled up with mud; a hole left for a frame serves as a window or a miserable door answers that purpose. The cracks in the mud are rarely filled up, and in very few instances only have I seen a house repaired. The floors are of clay, moist in itself and rendered more disagreeable by the filth of the inhabitants,

[7] *Loc. cit.*, p. 192.

with whom the pigs not infrequently dispute the right of possession." He goes on to describe the scanty furniture of the house, the inadequate clothing of the family, "and nothing can be more frugal than the whole economy of the table. So intent is the owner on employing his slaves solely in employment directly lucrative that the garden, on which almost the entire subsidence of the family depends, is kept in the most miserable disorder." . . . "Few, very few of the numerous class of miners from which the above instance is selected are rich, few are even comfortable; how wretched then must be the state of those who possess only eight or ten Negroes, or whose property does not exceed three or four hundred pounds."

These paragraphs need to be read in connection with his description of Ouro Preto, where he said the houses of the higher classes were "much more convenient and better furnished than any I saw in Rio de Janeiro and São Paulo." He was especially impressed with the carved beds, satin coverlets, and lace-trimmed pillows. There is no inconsistency here; it is characteristic of regions where low-grade people do all the work that a small fraction of the population is able to live in luxury and officials of various kinds make out fairly well, but the "poor whites" are in a parlous state, often lacking both the energy and intelligence to improve their situation by their own efforts, even if the organization of the social system permitted. Slave labor is characteristically inefficient, especially at tasks which require more than the simplest intelligence, because it has no incentive to do anything but implicitly obey. But the key to the situation is in Mawe's observation that when he questioned technique he was referred to a Negro for the answer. Not only was the labor delegated to the low-grade man, but also the understanding of the rationale of labor; the upper classes not only had abdicated from the labor of personal effort but from the labor of thinking as well. The only redeeming feature of slavery as an institution was, theoretically, that a large group controlled by a few best minds would accomplish more than the same number of individuals. In practice the "best minds" proved inefficient.

The inevitable outcome of such a social system was that, although the gold and diamond production of Minas in the eighteenth and early nineteenth centuries corresponded to a great deal of wealth, it had but little permanent effect in developing the region. The boom period did not yield high wages for the workers, to be spent locally, thus permitting the establishment of general business to supply them, nor was the "royal fifth" spent for local public improvements, such as schools or roads. Instead, it was used to support the extravagances of government at Rio de Janeiro and Lisbon, while the large incomes of the few were spent for luxurious living instead of invested in productive enterprises.

At the time of Mawe's visit the mining of diamonds was a state enterprise and had been for some thirty years. Gold production was still a private enterprise and he speaks of a code of laws under which discoverers of new deposits claimed a plot of ground, but gives no details. All the gold collected in the washing operations had to be delivered to the mint at Ouro Preto, where one-fifth was taken for the government and the rest melted into an ingot upon which the assay master stamped its weight, fineness, the place and date. The ingot was numbered and registered and a copy of the registry given to the owner with the ingot which he was then free to transport. He mentions the *garimpeiros,* who prospect the country but who do not file any notice of their discoveries, or make any return of the royal fifth, and himself visited a place only 160 miles from Rio de Janeiro where they had carried on their operations for three years before being discovered and for two years more before being brought under official control. Although the guards at the registry stations along the main roads searched suspicious travelers, confiscating gold that had not been melted into ingots and treating its possessors as smugglers, not a little of the gold produced must have thus escaped record.

Diamond mining was carried on differently. In 1729, the government suppressed gold mining in the diamond district (which adjoined the principal gold regions on the north) and allowed planters to put their slaves at diamond washing at an annual tax of $2.50 per slave. This resulted in so large a production that the tax was at first increased and eventually, in 1740, a few contractors were given a monopoly of diamond washing at an annual tax per slave that continued to increase until it finally reached $260 per worker per year. There was little or no profit in the industry at this figure but it greatly stimulated the activities of the *garimpeiros*. In 1772, the government took over the mining operations. All the best stones were sent to the royal treasury, but the cost of government supervision was so great that it is generally considered that the operations were thereafter generally unprofitable. Mawe computed[8] the average operating cost for the years 1801–1806, and estimated it at a little over $8 per carat, or more than the average selling price. The output he estimated as 20,000 carats annually. The gravel was dug out by hand and carried to sorting floors where it was washed and the slaves picked out the stones by hand. It is hinted that the slaves sometimes delivered their findings to the overseers privately, though precautions were

[8] *Loc. cit.*, p. 352.

taken against their securing any for their own private benefit. Mawe said he "found that diamonds were bartered for everything, and were actually much more current than specie." The output from the individual workings was sent to headquarters at Diamantina, thence under guard to Rio de Janeiro, and shipped to Lisbon, where they were sold through brokers to the cutters in Holland. After 1801, most of the output seems to have gone to London on account of loans made by Baring Brothers to the government of Portugal. After Brazil became an independent country, the government surrendered its operation of the Minas Geraes deposits to private initiative.

India was the principal source of diamonds at the time of their discovery in Brazil, and when Brazilian stones came on the market efforts were at first made to slur their quality, though they were actually of good color and quality. Some stones were at first sent to Europe by way of India. Later the good quality of Brazilian stones was admitted. Most of those found in the Diamantina district were less than 10 carats in weight, though a few of over fifty carats were found, and a slave who found one weighing over seventeen and one-half carats was given his freedom. A few years before Mawe's visit a stone weighing 1·44 carats was found in the Abaete river district. This is one of the seven regions in Minas in which diamonds have been found. The Grão Mogol district lies about 120 miles north of Diamantina; and somewhat farther distant, to the southwest, on the Bagagem river, the largest gem diamond found in Brazil, over 250 carats, was discovered in 1853. The first diamonds were found in gravels in the streams, but in 1834 one was found in clay at a higher level and systematic search has revealed a good many stones in hitherto unsuspected places, though in the same general region as the first discoveries.

About 1840, diamonds were found in Bahia and are now known to exist over an extensive area at the headwaters of the Paragassú river, near the center of the state. There are other regions in the state, but this is of most importance. The locality is remarkable, because many dark-colored stones, of amorphous structure, are found here, some of them of unusual size. One is said to have weighed 3,150 carats. At first they were supposed to be valueless, but about 1860 it was learned that they gave better service for use as the cutting stones in diamond drills than the crystallized stones, being less likely to fracture. Of recent years washing operations have had for their principal objective these carbonadoes, as they are called. Because of their scarcity and the commercial demand for them for use in core-drilling, polishing, cutting and wire-drawing, their price steadily rose until it considerably exceeded that of gem stones.

Up to the time of the World War most of the product was the result of small-scale panning operations; it was sold through brokers who dealt with Europe. The war greatly interfered with this trade and after the war it was difficult to secure enough carbonadoes, which reacted to increase the price. The Bahia Corporation was organized, secured a concession to a considerable area, and used modern equipment for prospecting and drag-line scrapers to handle the gravel. Owing to the character of the bed-rock, the scrapers proved unsatisfactory, and washing with hydraulic giants has been substituted. It is reported that this method has reduced the cost of operation to one-fifth of that using hand labor.

Meanwhile, the core-drill operators had been experimenting with the use of large numbers of small, cheap, off-color gem stones for setting their bits, in place of a few large carbonadoes, which had been the usual practice. This new development is reported to be successful and it seems probable that the demand for carbonadoes will considerably decline in the future, which would, of course, lead to a corresponding reduction in the price, since the supply of small off-color gem stones greatly exceeds the demand and they are therefore offered at a low price.

Diamonds are found in other parts of Brazil in addition to those mentioned. Some have been produced in the state of Goyaz, which adjoins Minas Geraes on the west, but little definite information is available regarding the deposits there. Far to the west in Matto Grosso, near Cuyubá, is another district. It is claimed that a million carats weight of diamonds were produced there prior to 1850, but of recent years production has been spasmodic, reflecting discoveries from time to time. A few diamonds have been found in the states of Parana, Piauhy, and São Paulo, but none of these places is an important center of production.

In Minas Geraes, production still continues, and from time to time companies backed by foreign capital have attempted to introduce large-scale methods of recovery. These have been almost uniformly unsuccessful. The present output chiefly results from small-scale hand work and the annual yield of diamonds in Brazil is mostly from Minas Geraes and Bahia. No figures are available as to the quantity and value of the output, but those in a position to judge estimate it at one to one and one-half million dollars worth in a normal year. The depression that began in 1929 had such a serious effect on gem diamonds that the owners of the South African mines, which now supply about seven-eights of the world's supply of diamonds, announced, in April, 1932, that they would shut down their properties for two years. Presumably the effect on Brazilian production will be correspondingly serious, and no one can forecast what the situation will be ten years from now.

Previous to 1800, all the diamonds found in Brazil were sent to Europe for cutting, but about that time cutting operations began in Rio de Janeiro, and are now carried on at various places. No one knows what proportion of the cut stones now sold to Brazilians and to tourists in the coast ports has been locally cut, but it has some importance at least. The local lapidaries are inferior in skill to those of Amsterdam and New York and consequently have to sell their product at lower prices.

Gold

All the gold-mining operations up to 1800 seem to have been the washing of surface deposits; Mawe makes no reference to any underground mine. No one has ever discovered the original source of the diamonds found in the existing streams and their ancient channels, but the source of the gold, it seems evident, was narrow gold-bearing veins in the older rocks, that had been concentrated by the processes of erosion. Among the rocks of presumably Algonkian age is a curious formation known as itabirite that is an intimate mixture of fine quartz sand and iron oxide, either finely laminated or massive. In places this contains narrow veins of native gold, or of gold-bearing quartz, giving rise to what are known as *jacutinga* ores, which are peculiar to Minas Geraes. Sometimes they are so soft that they can be washed with a stream of water. The first ones worked were probably secondary deposits, but the Gongo Socco mine, eighteen miles east of Sabará, which was reputed to have yielded eleven tons of gold between 1826 and 1829, the Itabira mine, and the Maquiné were in jacutinga deposits. All of them have long since been exhausted and at the present time (1934) only one important gold-mining company, the St. John del Rey, is operating underground in Minas Geraes.

This enterprise deserves description at some length, not only because it is the only surviving gold mine of importance in Brazil, but because it vies with another in South Africa as the deepest mine in the world, and also has some other unusual features. The company which operates it is the oldest registered English mining company, having been organized in 1830 to operate a gold mine at St. Joao del Rey (also in Minas Geraes). Work there proving unprofitable, operations were transferred to the present location in 1834, where they have been since continued. The ore deposit is remarkable in many ways. It has an unusual composition, consisting of quartz, carbonate of iron and magnesia (ankerite), albite, and sulphide minerals, the latter being about half, by weight. Of these pyrrhotite and arsenopyrite are the most important, though there is a small amount of chalcopyrite. The gold, which ordinarily amounts to $10 to $15 per short ton of ore, is apparently associated with the arsenopyrite.

The ore deposit has the shape of an enormous lima-bean pod, dipping into the earth at an angle of about 45° near the surface, but gradually flattening until it dips about 20° on the lowest levels, where it is also crumpled and folded. The thickness of the pod varies, reaching about thirty-five feet as a maximum, but typically being ten to fifteen feet, its area in cross-section ranging from 4,000 to 9,000 square feet. It has been followed along its axis for over two and a half miles by a series of vertical shafts and horizontal drifts; the lowest workings were reported, on May 1, 1933, as being 8,050 feet vertically below the surface opening. Writing of this mine in 1905, when its lowest workings were 3,700 feet deep, J. H. Curle said,[9] "The average yield of the ore appears to be gradually getting less with depth; it is now about forty shillings a long ton." But the report of the company for 1925 indicates that the yield then was about fifty-six shillings per long ton, the qualification "about" being necessary because in that year some two and one-half per cent of the ore milled came from another ore body, and no indication is given as to the relative gold content of this ore. It is also uncertain whether Curle included silver in the "yield" he spoke of. The silver content is small, only about 1 per cent of the total value, but since its price varies it affects precise statements of the value of the ore. The yield varies, having been in 1922 over sixty-two shillings per long ton, and in 1930 over fifty-one shillings, but there is no good evidence implying a gradual decline in gold content with increasing depth. The mine is entitled to outstanding rank both for the remarkable persistence of its ore body, and the absence of material change in gold content in depth. When and if the mine has to cease working, it will undoubtedly be because the increasing difficulty and cost of working at such great depths makes it no longer profitable, not because the ore has given out.[10]

That it would be increasingly difficult to extract ore

[9] J. H. Curle: "Gold Mines of the World," 2nd Ed., p. 278, 1905.

[10] The main vein was quite low-grade on the 7,126-foot level, but exploration revealed another ore body of good size and value. This was also lower in grade on the 7,626-foot level; both veins are now being explored in greater depth, and the 1931 report of the company stated that over 1,500,000 tons of ore of workable grade are known to exist. This apparently is of distinctly lower grade than prevailed above the 6,926-foot level.

from depths that are more than one and one-half miles below the surface of the earth might be taken for granted, but some further explanation of it seems worth while. Between 1837 and 1868, the deposit was worked practically as an open cut, but in 1874 shafts were sunk, leaving pillars of ore beneath the old workings, and extraction was continued, supporting the open stopes, where the ore had been taken out, with timber. By 1886, though altogether something like $1,500,000 worth of timber had been used for this purpose, the workings collapsed and the company was about to abandon the mine. The then manager, G. Chalmers, fortunately persuaded the directors that the deeper ore could be worked by placing the shafts in solid rock away from the ore body, and filling the openings with material brought from the surface as soon as practicable after the ore was taken out. Something like three-quarters of the total yield of the mine has been obtained since this plan was adopted, so it was eminently well advised. But it involved handling surface materials into the workings at the great depths reached, as well as taking ore out, so the effects of increasing depth are correspondingly more potent.

Since the ore body dips it has been followed by a series of vertical shafts and horizontal drifts from the point at which it was intersected in 1886. Cars going into the mine first pass through a horizontal drift for about 1,000 feet and then are sent down a shaft 1,940 feet. Next they traverse a horizontal drift for 600 feet and then 1,160 feet down another shaft, followed by 1,400 feet of travel on the level and 1,200 feet down another shaft. After 1,400 feet more travel on the level, a further descent of 1,200 feet, and 2,100 feet of travel on the 5,826-foot level, the top of the "H" shaft is reached. At the bottom of the "H" shaft there is a horizontal drift about 800 feet long to the top of the "I" shaft, which is inclined at 17° to the horizontal and has a vertical depth of 300 feet and a length on the incline of 1,030 feet. The bottom of the "I" shaft is connected by a horizontal drift about 800 feet long to the top of the "J" shaft, which is also inclined at 17° and has a vertical depth of 400 feet and a length on the incline of 1,370 feet. The "J" shaft bottom is at the 7,126-foot level and the lower levels are reached by subsidiary shafts and inclines. The handling involved in these changes from horizontal or inclined, to vertical movement and back again, adds to the expense of working, and the possibility of connecting the deep workings with the surface through a single shaft has been considered, but has so far not been attempted on account of the capital expenditure involved.

This is not the only disadvantage of working in depth, however. With increase of depth below the surface, the earth's temperature increases. The rock temperature here at a depth of 6,726 feet was over 121° F. Since rock is a poor conductor of heat, this condition could be met by circulating sufficient quantities of outside air, which there ordinarily has a monthly average temperature of from 62° F. to 73° F., were it not for the temperature effects of pressure on air. When air descends into a mine it increases about 1° F. in temperature with each 300 feet in depth, unless it is cooled by taking up of moisture from the wet walls. This mine is so deep and so completely dry below the 4,600-foot level that the surface air would not only be uncomfortably warm upon reaching the bottom but would have but little effect in reducing the rock temperature.

In 1920, the dry-bulb temperature on the 5,826-foot level was 101° F. and the wet-bulb 87° F. One does not require a detailed understanding of the physiological effects of high temperature and humidity on human efficiency to understand that under such conditions the output of labor would be low. At the end of 1920, an air-cooling and drying plant, which cost about $450,000, was put in operation on the surface. In 1930, the good effects of this were supplemented by an ethyl-dichloride air-cooling plant constructed on the 5,826-foot level. These plants have reduced the effective temperature on the lowest levels to such a degree that work can continue but they, of course, add somewhat to the working cost. On the other hand, working costs in 1930 were brought down to 36s 9¼d per long ton, as compared to an average of 43s 1¼d per long ton, for the ten preceding years, chiefly by increasing the tonnage handled by one-third. Thus operating costs have actually been lowered in the face of increasing physical difficulties. But just as even Methuselah eventually died, so the working of this ore body, which has been continuous over a period of 100 years, must eventually come to an end.

That, however, does not mean that underground gold production in Minas Geraes will completely cease, for the company has, for ten years or so, been developing nearby another underground ore body. The ore in this is lower grade, but will doubtless serve to keep part, at least, of the existing plant in operation after work on the original ore body ceases.

The following notes on general conditions at the mine were obtained in the course of a visit[11] there in 1922. Adjacent to the mine has grown up the town of Villa Nova de Lima, with a population of 12,000 to 14,000, of whom about 300 are European employees (and their families) of the mining company. The total number of employees of the mining company at all the places where it has

[11] T.T.R.

work going on is between 3,000 and 4,000, of whom about 150 are English. The company has built practically all of the living accommodations in the town, a hospital, a dairy, a slaughter-house, refrigerating plant, and retail meat shop, a small works for bottling soft drinks, and a hall where moving pictures are shown weekly. For the English employees there are a church and a school, club house, tennis court, and cricket field, all provided by the company. During the past few years the company has built for the Brazilian workmen and their families 150 houses, each with four rooms, kitchen, lavatory and shower-bath; more of these are to be built in the near future. In 1930, the company spent £6,864 on miners' houses.

The older living quarters for Brazilian workmen are long, one-story buildings, divided into two-room segments; the front room for cooking and living and the rear one for sleeping. Each is about ten feet square and the front one has a window and door, and the back one a window, with a board shutter instead of glass. Since the temperature range is seldom much below 60° F. or much above 75°, so simple a dwelling is adequate to meet the climatic conditions. All the washing, including the dishes, is done at public washing places, where flat trays and running water have been provided. Water for domestic purposes is carried in a bucket from public fountains, the spaces between the living quarters are lighted with electric lights, and there are public latrines at convenient intervals, though the children do not always use them. The rent for such quarters is two or three milreis (say twenty-five to forty cents) per month. Even such simple and rather primitive quarters are better than the general standard of living of the working people of the country. One of the problems of operation is to maintain a sufficient supply of labor and the company does not feel in a position to enforce a high standard of living. The men who work in the stopes and on development work are paid on a contract basis and could raise their scale of living through increasing their income by increased efficiency in their work if they felt any need for it. The tons produced per man per day employed underground are low, only a fraction of what it would be in a corresponding mine in the United States. While the rate of pay of seven to eight milreis per man per day (say $1 gold in 1922) seems low to one accustomed to thinking only in terms of wage scales, it is not low in proportion to the amount of productive work done.

One of the important social functions of the mining company is the provision of an adequate hospital and systematic care for the health of the workers. How great the need for this is, is indicated by unofficial statements that in places 95 to 100 per cent of the rural population is infected with hookworm, and malaria is common enough to be a social problem. The official statistics exhibited by the Department of Public Health at the Exposition at Rio de Janeiro in 1922 showed a high percentage of infection of all kinds of venereal diseases. Tuberculosis is a serious problem even in the temperate climate of the high plateau; it seems to be promoted by a superstitious fear of night air, which leads to numbers of persons sleeping in the same room with the windows closed. The dietetic habits of the common people also do not seem to be good. They eat little or no bread, their carbohydrate intake being derived from rice, "mandioca," or "farinha," beans, and "angu," a sort of pudding made of ground corn. There is general agreement that manual workers ought to eat a hearty breakfast, have if possible a warm heavy meal in the middle of the day, and take a light evening meal. The Brazilian miner, however, goes to work in the morning having had practically nothing but a cup of coffee on rising, eats a light lunch about 10:30 a. m. and has his heaviest meal in the evening.

Something more must be said about climate. The average temperature at the St. John del Rey mine is 68° F. with monthly ranges of from 62° F. to 72½° F. Norfolk, Va., which we think of as having a southern climate, has a mean temperature for the year of 60° F. with a mean for the hottest month of 79° F. and for the coolest month of 41° F. In other words, the coolest month at Norfolk is 20° cooler than at St. John del Rey, and the warmest month 6° warmer, while the average for the year is 8° lower. Students of the effect of climate on man would probably consider that the climate at St. John del Rey, in being warmer and especially less subject to change than that at Norfolk, was therefore less favorable to human energy and health.

It is not surprising, therefore, to find the efficiency of mine workers in Brazil low. A somewhat unfavorable climate, habits of diet and living that are not conducive to the best health, and a high incidence of infection and disease are added handicaps to the considerable percentage of Negro blood in the ordinary workman. If their effects are visible to a critical and inquiring visitor to a mining development that has been in continuous operation for a hundred years, it can be readily appreciated that they must be taken into account in planning any new development in a region where the labor supply would be predominantly agricultural and exhibit these handicaps in enhanced degree.

These remarks apply to health conditions on the central plateau. Dr. R. J. Needles, of the Hospital of Cia Ford Industrial do Brazil, at Boa Vista, in the Amazon valley, has recently published[12] results of studies on the natives of that region. These indicate that 50 to 75 per cent of the natives have chronic malaria, 90 per cent have intestinal parasites, more than 90 per cent show well-marked secondary anemia, a large percentage have chronically enlarged spleens, and malnutrition is characteristic, the diet not only lacking in variety but being insufficient in amount. The result is a condition of general lethargy. The seriousness of this as a factor in the future development of Brazil needs no emphasis.

Before leaving the operations of the St. John del Rey, two of its handicaps must be referred to since they would apply equally to any other enterprise in that region. The first is the shortage of timber supply. This seems incredible to one who thinks in terms of pictures of Amazon forests, but this region is far distant from the Amazon, and the plateau actually has only a scanty growth of trees. It is made worse by the national habit of cutting down brush at the beginning of the dry season, allowing it to dry, and then setting the area on fire. The aim of this is to remove the brush, permitting grass to grow for pasture purposes; it not only has an inimical effect on the physical texture of the top soil, but prevents the attaining of maturity by second-growth timber. Also the quick-growing soft coniferous woods that are in almost universal use in the United States are completely absent from this region. Many kinds of hard wood that are completely unknown in the northern hemisphere are utilized, especially aroeira and peroba; being hard wood they are relatively difficult to work, and having to be brought from a distance, are correspondingly expensive. The St. John del Rey company carried on for a long time an attempt to raise eucalyptus trees as a source of timber

[12] *Science*, vol. 78, p. 532, 1933.

supply but finally gave it up. Several railroads in the plateau region in Brazil have also attempted the growing of eucalyptus to provide them with cross ties, but this has also been generally unsuccessful. The timber problem is particularly acute at the St. John del Rey because it has been operating so long that local sources of supply have been exhausted.

Another problem is that of power supply. The almost complete absence of coal in Brazil will be discussed later, and what has just been said as to timber will explain why wood-burning power plants are not practicable. In a highly dissected plateau region one would naturally expect that hydro-electric power would be easily available, but since the sixty-four inches average rainfall all occurs between November 1 and March 30, a corresponding investment in dams and storage reservoirs is required. Between 1906 and 1922 the St. John del Rey developed nine hydro-electric power sites, but the total capacity of all nine amounted to only 5,545 kva., or, say, something less than 6,000 horse-power. There are large waterfalls in the plateau region, and if a mine fortunately was near one it would have ample power supply; a large project with ample financial backing could bear the cost of a long transmission line, but prospects and small mines would be handicapped. In California the long transmission lines have been financed by the market for power in large cities and in the oil-fields; in many parts of the state it is possible to connect with them at relatively small expense. In Minas Geraes there are no oil-fields or any other form of industry that requires power in large amounts and the largest city (Bello Horizonte) has a population of only about 110,000. It is a distributing center for an agricultural area and the power required for lighting and street railways is obtained from a 13,200 kva. station at Rio des Pedras, at no great distance.

Of gold production in the states of Brazil other than Minas Geraes relatively little need be said. Careless reading of what has been published regarding them is likely to leave the impression that they are regions of wonderful promise. A Brazilian writer says of the state of Matto Grosso, which has an area of almost four times the size of Minas Geraes and only one-eighth of its population, that gold can be found in "innumerable" streams throughout the state. One must set such a statement against the cold fact that gold was discovered in Matto Grosso more than 200 years ago, production at no time was very large, and at present is less than it was formerly. Such gold as has been obtained has been derived chiefly from the gravels of existing streams or scattered patches of Quaternary gravels. Many attempts have been made to work gold-bearing lodes, but none of the mines has attained much depth or much size. Precise information regarding the gold possibilities of the state does not seem to be available, but since the possibility of using dredges to work low-grade gravels has been discussed for more than twenty years without any actual development resulting, it seems reasonable to infer that those who have done exploratory work, but have not published their results, found conditions discouraging. No one can say that important new discoveries will not be made, but after two centuries of exploration they at least do not seem probable. In São Paulo a new gold mine, the St. George Mining Co., at Araçaiguama, has recently been opened, but no details regarding it are available.

In the state of Bahia some gold has been obtained in connection with diamond production and attempts to work gold-bearing veins have been made near Jacobina and elsewhere. A recent statement by H. S. Williams, long a member of the Geological Survey of Brazil, is as follows:

"Gold is washed in various streams of the Serra de Assuruá on the northwestern side of the Chapada Diamantina in a haphazard manner, and frequent remittances of gold and nuggets are made to the city of Bahia. About forty years ago a huge nugget weighing six pounds was found near the village of Gentio de Ouro in the Serra de Assuruá, and was shown in Rio de Janeiro, where on the strength of this one nugget a company was formed and capital realized. An expedition was equipped and proceeded to the locality. After adequate study it was determined that the nearest water supply was to be found in the Serra de Burity Quebrado forty kilometers distant to the south. The necessary material was ordered and finally delivered with infinite difficulty, and an iron pipe-line forty-odd kilometers long was completed and water actually delivered at Gentio, but this effort exhausted the capital of the company and all efforts to raise capital for operation failed, so the project was abandoned. Gold is found at Mamonas near Cascavel, near Martin Vaz, and at Jacobina, where in the olden days considerable gold was mined in alluvial washings.

. . . "About the city of Minas de Rio das Contas, gold washing is practised continually by the inhabitants, apparently with fair results, but no attempt has ever been made to determine the origin of this gold or to mine the region on a systematic basis."

It must be remembered in this connection that only along its southeastern coast does Bahia have a heavy rainfall. In a dry year about half the area of the state receives so little rain as to rate as a desert, and in normal years much of the interior would be classed as a region of small rainfall. As the working of surface deposits of gold usually involves free use of water, the natural conditions are therefore somewhat unfavorable, but if any large and valuable gold deposits had been found in the state means would have been found to cope with unfavorable conditions. It seems improbable, therefore, that gold production there will much exceed the small amounts that have long been produced.

In Goyaz, some gold has been produced for the past 100 years; a typical reference to it is that "the production has no doubt been considerable." Both placers and lodes have been worked, but the operations have never attained more than a small scale.

About the only definite information regarding gold in the state of Rio Grande do Sul occurs in an article by H. Kilburn Scott. Writing in 1904, he said,[13] "In São Sepe, twenty-two miles northwest of Cacapava, auriferous quartz veins are plentiful over an area twelve and one-half miles long by four to six miles broad. . . . Near the village of Dom Pedrito thirty-seven miles to the southwest of Lavras, the Barcellos Gold Mining Company worked some auriferous deposits about fifteen years ago (1888)." This sounds encouraging, but by 1932 production in the district had not developed to the point where there were any definite figures on it.

In that part of Pará which borders on French Guiana, gold was discovered in 1893 and some 6,000 men went into the region about Calcoene. Ten years later less than 100 remained, and now only a little gold produced by the natives comes from there, the lode workings having been abandoned.

As to Ceara, Piauhy, and Maranhão, H. C. Carr spent

[13] H. Kilburn Scott: "Mineral Resources of Rio Grande do Sul," *Trans. I. M. E.*, vol. 25, p. 510, London, 1904.

six months exploring these regions in 1913 and his report on them may be summed up in the following quotation:[14] "Northern Brazil, in my estimation, does not hold out any inducements to the prospector, with the possible exception of the Turyassu [river] and the limestone hills in northern Piauhy." It is interesting to note that this mining engineer thinks unfavorably of the mineral prospects, but is otherwise optimistic, as he refers to rubber, hardwood, cattle, and cacao and says that "anyone with a small amount of capital and a large amount of industry can easily accumulate a fortune there."

The figures for the gold production of Brazil given by E. P. de Oliveira, Director of the Servicio Geologico do Brasil, in "Gold Resources of the World," published at Pretoria in 1930, indicate that the yearly output has ranged between 100,000 and 150,000 ounces for the past twenty years. The St. John del Rey has recently been yielding about 90 per cent of the total gold; that mine, and the Ouro Preto Gold Mines of Brazil which produced about 25,000 ounces yearly until 1927, have indeed furnished three-fifths of the total output of gold in Brazil in the past century. If, as, and when the St. John del Rey ceases operations the gold output of Brazil will undoubtedly decline to a very small fraction of its present output of one-half of 1 per cent of the world's output.

This review of gold and diamond mining in Minas Geraes leads up to the question why the impetus it gave to immigration into the region did not produce a permanent economic development comparable to that which resulted from the discovery of gold in California 150 years later. There seems to be no single answer to the

[14] H. C. Carr: "Mining Reconnaissance of Northern Brazil," *Eng. and Min. Jour.*, p. 399, August 29, 1914.

question, for many factors are involved. Minas Geraes is not so well adapted to agricultural production as California, and the high cost of transportation of agricultural products which its physical situation entails would handicap even a very rich region. If a single factor must be cited the choice would fall on the difference in the labor characteristics of the two regions. California was peopled by adventurous white men of every occupation who went there intending to produce gold by their own labor. Many of them promptly turned to the occupations with which they were more familiar, and they all worked. But gold and diamonds in Brazil were produced by slaves—the white man did not work. If his venture was successful, he tended to leave the country to expend his gains, if unsuccessful he probably also left. John Mawe's account of the region in 1810 (see p. 103) gives more details. Agriculture and general industry, therefore, developed only slowly and the conditions were less favorable to their development. Progress, though slow, now seems to be steady.

Coal

Coal was never a mineral of much importance until steam engines began to be applied for steamboat and railway transportation, and the demand this created for more iron reacted in turn to increase its use in ironmaking. It is not surprising, therefore, that coal was of little interest in the early history of Brazil; the more so because, in a country lying so near the equator and settled by a people who came from a region where, broadly speaking, it is not the custom to heat dwellings even at altitudes and at times of year when North Americans would consider heat necessary, it was natural that there would be little demand for fuel except for cooking purposes. This demand was easily met by wood. When railroad construction began in Brazil in 1854, and subsequently when demands for power for general purposes began to increase, interest in a domestic coal supply began to awaken.

From what has been said about the general geology of Brazil, an economic geologist would at once realize that over much of its area coal deposits of importance can scarcely be expected to occur. But from Southern São Paulo to the edge of Uruguay geological conditions are more favorable. The Rio Bonito beds of the Tuberão series (Lower Permian) contain some coal almost everywhere they occur in Brazil, and in the states of Santa Catharina and Rio Grande do Sul the coal is thick enough to permit being worked. A great deal of effort, both in exploration and exploitation, has been expended on these deposits. About thirty years ago, Dr. I. C. White, then state geologist of West Virginia, and a leading American authority on coal geology, was retained as the chief of a commission to study the coal resources of Brazil. It is not necessary to review the work of this commission in detail since it may safely be assumed that its studies were thorough and complete, while the following quotation from the final report of the commission (Rio de Janeiro, 1908) succinctly states its conclusions.

"The result of the work of the Coal Commission during the year 1904 (which was fully confirmed by the further studies in 1905–1906) was to abandon hope of finding beds of pure coal of exploitable thickness in Brazil, since it was only too apparent at every outcrop and exploitation visited that the coal was everywhere of practically the same quality; that while some fairly pure coal existed in every bed, it was so interstratified and ingrained with bituminous shale or slate as to render the separation by ordinary mining methods impossible."

Perhaps the general reader may not immediately perceive that this statement does not say that coal cannot be mined in Brazil; it says that by ordinary mining methods, which suffice to produce clean coal in other coal-producing countries, clean coal of good quality cannot be produced there. From coal which is too high in ash and sulphur to be usable as mined (as is the case with Brazilian coal) a usable coal can be produced by purification processes. But naturally clean coal of excellent quality usually sells for less than one-tenth cent per pound at the mine, in the United States, so it is instantly apparent that if perhaps a third of the coal as mined has to be rejected in a cleaning process the cost of the product would be so high as to make it impracticable to compete with the cheaper natural product.

Brazilian coal, of course, has an advantage in that imported coal from Europe and the United States has to be brought many thousands of miles, but ocean freights, especially on the type of steamship which transports coal, are quite low. And unfortunately the coal deposits in Brazil are not near any principal center of coal consumption. Their product after cleaning requires transportation to ports, coastwise transportation, and the same amount of handling thereafter as good quality imported coal. Consequently the Brazilian coal producers find great difficulty in marketing their cleaned coal in competition with naturally clean imported coal, even with the aid of government support, extended in various ways, and the patriotic sentiments that conduce to its use.

The place which domestic coal production has attained in the fuel supply of Brazil can be inferred from the fact that after nearly a half-century of attempts to develop it, the highest year's output claimed was 500,000 tons of coal averaging 20 to 30 per cent ash, while the giving of the figures only to the nearest hundred thousand renders them suspect. After washing to lower the ash content it is used as fuel on the government-owned railways, the fire-boxes being specially constructed to permit burning material of that size and composition. Coal imports since 1920 are shown in the following table:

Coal Production and Imports—Brazil, Tons

	Imports	Production
1920	1,120,575	300,000
1921	843,132	400,000
1922	1,176,287	500,000
1923	1,469,756	324,000
1924	1,613,578	342,200
1925	1,702,823	392,376
1926	1,771,858	400,000
1927	1,976,000	400,000
1928	1,919,000	400,000
1929	2,035,000	400,000
1930	1,890,000	400,000
1931*	1,265,000	461,500
1932*	1,171,000	322,131

* Includes coke and patent fuel.

It is evident from this that domestic coal resources have not contributed any important part of Brazil's fuel supply in the past decade. The question whether they can, in the future, be expected to attain a more important place is inseparable from the question as to whether it is desirable for the country's welfare that they should. There are, of course, many persons in Brazil who are quite firmly convinced that every possible means should be employed to increase domestic coal production; their reasons for holding such a view may easily be inferred. On the other hand, such means will necessarily involve either subsidizing the local industry by an import tariff to raise the price of imported fuel as delivered in Brazil to a level at which domestic producers can compete, or through some form of direct government support. The present requiring of the railroads to use domestic fuel is practicable because they are government-owned, but it is clear that this simply amounts to transferring to the coal producers the benefit that would accrue to the railroads if they were permitted to use the best quality of fuel obtainable at the lowest price. Commercial consumers could not be so coerced, and the coal mines would have to receive some direct subvention from the government. It is not a case of affording temporary support to an "infant industry" to enable it to establish itself in the face of foreign competition, as at first hoped, but of providing continuing assistance to overcome the handicap of natural conditions. This assistance would have to be derived either from general tax receipts, or from a tariff on fuel that is, for practicable purposes, a special tax on coal consumers. It may fairly be asked whether the general social good of Brazil would not be better served by freedom from such taxation, than it would by the artificial stimulation of a domestic coal industry.

Petroleum

Petroleum, as the second most important fuel mineral, naturally follows coal in its claim for consideration. Since the petroleum industry the world over is less than a century old, it was, of course, not a subject of interest in early Brazil. Its rapid development in the U. S. after 1870 was in large part contingent on the successful marketing in foreign lands of a large fraction of its refined products. While kerosene as a source of light has had to yield to the competition of electricity in the urban areas of Brazil, as elsewhere in the world, it is still of great importance in agricultural areas (the major part of Brazil) which are not grid-ironed with light and power transmission lines, as is the case in the United States. Automotive transportation is of course dependent on gasoline supply, and fuel oil has found a considerable market in the urban areas in competition with coal as a source of power. Since there is no domestic production as yet, the part which petroleum and its products play in the economic life of Brazil can be inferred from the following tables, which give data as to recent imports:

Petroleum Imports of Brazil

	Bbl.	Value
GASOLINE		
1927	1,732,000	$13,110,000
1928	2,189,000	14,061,000
1929	2,527,000	17,376,000
1930	2,405,000	14,905,000
1931	1,844,000	6,766,000
1932	1,237,000	3,839,000
KEROSENE		
1927	889,000	$ 6,801,000
1928	825,000	6,061,000
1929	933,000	6,852,000
1930	719,000	5,017,000
1931	784,000	4,230,000
1932	374,000	1,783,000
FUEL OIL		
1927	2,443,000	$ 6,043,000
1928	2,311,000	3,990,000
1929	2,296,000	4,071,000
1930	2,553,000	4,519,000
1931	2,673,000	4,100,000
1932	2,756,000	3,417,000
LUBRICATING OILS		
1927	255,796	$ 4,002,000
1928	288,186	4,102,000
1929	340,171	4,861,000
1930	184,176	

It is a question whether any domestic production of petroleum can be developed in Brazil. Unfortunately, the outlook is discouraging, for, geologically speaking, Brazil is largely too old. Careful students of its possibilities believe that only three areas have prospective value. These are the foothills of the Andes, part of the basin of the Paraná river, and a section of the Amazon basin where there are numerous slate outcrops. In addition to the uncertainty of the presence of oil, these areas are remote from transportation facilities, so that the large established petroleum companies, with world-wide commitments in prospective areas and present productive capacity in excess of existing markets have little incentive to venture large sums in exploration and development work. This is the more true because the political conditions that apply to the development of the iron industry, to be discussed in a moment, apply with equal force to the petroleum industry. In addition, the petroleum companies that are engaged in the import of petroleum products into and its distribution within Brazil have not enjoyed particularly happy relations with the government there. Were it not for the loss that would be incurred on their capital investments in storage and distribution facilities by withdrawing from the business, it is possible that some, at least, of them would consider it better business to cease attempting to market petroleum in Brazil rather than to continue under the conditions imposed on them.

Thus petroleum possibilities in Brazil face a dilemma. Capital is not available within the country to make the large investments and incur the large risks involved in attempting to develop an industry in the face of unfavorable conditions, while foreign capital is not likely to take the risks in the absence of a more favorable attitude toward it than now exists. The government of Brazil has been spending small sums on exploratory work, but the natural conditions call for large expenditures made in a bold way, and no government agency could hope to survive the political shock of expending such sums in ventures that prove unsuccessful, as large business organizations constantly do. No one can positively say that a domestic source of petroleum supply will not be developed in Brazil within the next half-century, but both natural and political economic conditions are unfavorable to the realization of such a hope.

In addition to search for petroleum deposits of the ordinary type, much attention has been given in Brazil to the possibility of the production of petroleum products from alternative sources. At various points around Camamaú bay, about sixty miles south and west of Bahia (city) a hydrocarbon deposit, to which Derby gave the name marahunite,[15] is found. It is reported to contain 70 per cent volatile combustible matter and 10 per cent fixed carbon. As early as 1859 Jose Nascimento obtained a concession to work this deposit, which he afterwards transferred to an English company. A plant of fifty Henderson retorts was completed in 1891 but in 1894 the operations of the company were shut down, and have not since been resumed; the capital investment made is reported as $2,500,000. It is claimed that the failure was due to mismanagement, but it seems strange that no subsequent attempts were made to operate, unless the results of the first three years were conclusively unfavorable.

In southern Brazil a formation known as the Iraty black shale occurs at various places, widely distributed from São Paulo to the Uruguay border. At various times in the past twenty years much attention has been given to the possibility of production of shale oil from these deposits by distillation, without any success as yet. At pres-

[15] It is also frequently referred to as *turfa*.

ent world prices of petroleum, production of oil from shale has but little commercial possibility anywhere. How the situation may change in the future no one can certainly foresee.

Iron

The existence of iron ore in Brazil was early noted. An exploring party in 1590 reported it in a mountain about sixty miles south of São Paulo. In 1597, officials were sent out from Portugal to promote mining in the region and brought an iron founder with them. One or two small forges were set up at Ipanema and probably produced some iron about 1600, making it the earliest iron-producing enterprise in the New World. Operations there were apparently discontinued about 1630, not resumed until 1765, and again discontinued after a few years. A blast furnace was built in 1800, but failed. In 1810, the government had four direct-process furnaces constructed by a Swedish metallurgist, and in 1818 two small blast furnaces were built. They must have been very small as they were credited only with a daily production of three to four tons. The ore was magnetite, and not easy to work, for it is high in phosphorus and titanium.

Meanwhile, small amounts of iron were being made in various parts of Brazil. Iron ore can be found in almost every state, and any clever blacksmith can reduce ore either in his forge or in a small primitive furnace that is easily made. Eschwege, writing of Brazil in 1811, said that most of the smithies he visited produced their own iron; some of them actually supposed they had invented the process, which is almost as old as human civilization. Eschwege showed them how to improve their methods, and started a plant near Ouro Preto in 1812 that could produce about 100 pounds of iron daily.

Small furnaces of this primitive sort began to spring up in various parts of Minas Geraes, and an observer in 1864 estimated that there were some 120 of them. O. A. Derby, writing in 1909, said that many of them "are still in operation."

It must be recalled that civilization as developed by 1800 did not require much iron, and it was the development of railroads shortly after that time that stimulated the development of steam engines and required a great increase in the output. Brazil's first railroad dates from only 1854, so its requirements in iron up to that time must have been comparatively small, and the economic struggle between iron articles of good quality imported from Europe and cheaper but inferior ones made by local artisans went on in Brazil without attracting much attention, since the total trade did not amount to much. When Brazil began to construct railways and power plants it was with the aid of foreign capital, and lacking any existing facilities to produce the things needed they were naturally imported from abroad. Up to 1888, aside from the primitive furnaces noted and the government-owned and operated plant at Ipanema, São Paulo, with a capacity of three to four tons daily, there was no iron production in Brazil. An attempt to start a small blast furnace near Morro de Pilar, south of Tejuco, Minas Geraes, shortly after 1810, had ended in failure.

In 1888, a small charcoal blast furnace, with a capacity of four tons daily, was erected at Esperança, near Itabira, in Minas Geraes. By 1900, it was claimed to have a capacity of ten tons daily (Derby said six tons daily in 1909) and had two five-ton cupolas, with facilities for casting pipe. It had good modern equipment, so its slow development must be attributed chiefly to commercial and managerial deficiencies rather than natural conditions other than the limited supply of wood suitable for making charcoal.

Iron-making began in Brazil about 1600 and in what is now the United States in 1644; in a small way in both places. By 1810, iron-making here had spread all over western Pennsylvania and adjoining regions; pig-iron production for that year is given as 54,000 tons. In Brazil, the government was then still attempting to promote the enterprise at Ipanema, but there was no other iron-making beyond the small quantities made by blacksmiths, as described by Eschwege, a total of perhaps 500 tons yearly. By 1928, pig-iron production in the United States had reached 38,000,000 tons, and steel production nearly 52,000,000 tons. No exact figures are available for iron production in Brazil that year but it probably did not exceed 25,000 tons of pig iron, and an even smaller quantity of steel. Brazil has even more good iron ore than is available in the United States; the question now arises as to why iron production did not develop there?

One clue has already been suggested. All the early iron was made with charcoal as fuel, and in Brazil the wood supply was scanty at most places where settlements had developed and such timber as existed nearby was soon cut. Too much weight should not be attached to this, however, because the small quantities of iron produced at that time made a correspondingly small demand on the fuel supply. Quite as important is the fact that the kinds of wood available in Brazil are not desirable for charcoal-making, for which soft wood is preferred. But when full allowance is made for these handicaps it still would have been possible to develop some iron-making in Brazil if other conditions had been favorable to it.

The most significant of the unfavorable conditions was that no large demand for iron developed in Brazil. Agricultural production was carried on by methods as primitive as the type of laborer employed, and beyond a few simple tools, cooking utensils, and harness trappings the *fazendas* had little need of metal. Their simple requirements were met by importations from Europe. Even had the crude metal been available in Brazil, few metal workers were available who were possessed of sufficient skill to fabricate it as desired; the metal workers were not available because there was not enough business in most localities to afford them a living. It must be remembered that the people who came to the United States typically came to make it their permanent home; many of those who came to Brazil sought a fortune that would enable them to live better on returning to Europe. Aside from gold, metal production offered but little hope of any early or rich return.

When railroad construction began in Brazil, about a quarter-century later than in the United States, it did not stimulate iron production. In the United States there was an already developed iron industry which quickly expanded to provide the facilities necessary to meet the increased demand, but in Brazil there was no industry, and this, combined with the pressure to use foreign steel, which resulted from the obtaining of the capital for railroad building in Europe, was why we find no corresponding increase in iron production in the period of railroad building. Railroads are quantitatively less important in Brazil than in the United States; in a slightly larger area Brazil has only 19,000 miles of railroad as compared to 250,000 miles in the United States. These figures seem striking enough, but fail to set the full contrast, for in the United States there also exist over

100,000 miles of oil pipe-lines, nearly an equal length of gas lines, water-distribution lines of unknown length, and in Pennsylvania the mileage of railroad track underground in the mines is greater than that of ordinary railroads on the surface. None of these things exists in Brazil, while the demand for iron and steel in manufacturing plants and building construction is only a tiny fraction of the demand here. In the nineteenth century the Brazilian demand for iron and steel developed much more slowly than in the United States, but even at that the industry, instead of developing to meet such demand as existed,[16] almost completely failed to develop.

Up to 1910, there was but little interest in Brazil in the possibilities of an iron and steel industry there, and no interest at all in the outside world. The situation quickly changed, however. In 1905, A. E. Törnebohm made a report on the iron-ore supply of Sweden. Excerpts from this published in the daily press of other countries conveyed to the general public the impression that the world was likely to run short of iron ore within the next hundred years. Conservation of natural resources, then already much discussed, became a burning political question, and the alarm thus excited was only partly allayed when, in 1910, the International Geological Congress published a two-volume report on the iron-ore resources of the world that showed them to be much larger than had been supposed. The part of the report which attracted most attention, probably, was the section on Brazil, written by O. A. Derby, an American geologist who had been head of the Geological Survey

[16] Emilio Teixeira, in *Eng. and Min. Jour.*, p. 734, November 5, 1927, states that the iron and steel requirements of Brazil in 1913 were about 500,000 tons, and in 1923 about 300,000 tons. Actual imports in 1927 were about 132,000 tons, in 1930 50,000 tons, and in 1932 were 29,359 tons, not including manufactures of iron and steel.

there since 1886. He had written several previous papers on iron ore in Brazil that attracted but little attention. But his statement, in the volume just mentioned, that there existed in Minas Geraes over 6 billion tons of iron ore of good quality that would average about 50 per cent iron, and a good deal much higher in iron content, instantly attracted attention in the outside world, which reacted to create interest in Brazil.

Perhaps the best way to picture the effect of this announcement is to quote from a consular report made to the United States Department of Commerce in 1918.[17] This report stated that the Itabira Iron Ore Co., an English concern in which the Rothschilds, Baring Bros., and E. Sassel were interested, had bought the Esmeril and Conceião deposits, near Itabira de Matto Dentro, and estimated to contain 300,000,000 tons of ore, for $600,000. The Brazilian Iron & Steel Company, an American concern, had purchased deposits in the same general region, capable of producing about 150,000,000 tons of ore, for about $110,000. The Minas Geraes Iron Syndicate, another American group, also produced iron-ore lands, but neither area nor price was stated. The Société Franco-Brésilliénne bought another deposit, estimated to contain 10 million tons of ore, for $50,000. The Société Civil de Mines de Fer des Juganda, a French company, paid $25,000 for a deposit estimated to contain 15,000,000 tons. The Deutsche Luxemburgische Bergwerks und Hütten Aktiengesellschaft bought a deposit for $25,000. The Corrego de Meio deposits, near Sabará, were acquired, presumably for a German syndicate, for $112,000. The Bracuhy Falls Co. purchased iron-ore lands for about $42,000.

[17] See also E. C. Harder: "The Iron Industry in Brazil," *Trans. A. I. M. E.*, vol. 50, pp. 143-160, 1914.

BRAZIL

In Brazil, the Companhia Metallurgica Brazileira was organized, taking over lands, supposed to contain 100,-000,000 tons of ore. The Companhia Siderurgica Mineira was organized at Bello Horizonte, with a capital of $87,500 and took over iron deposits in Minas Geraes. The Companhia de Mineração Metallurgia do Brazil was organized in Rio de Janeiro to take over the iron-ore lands of Antonio Lage. Carlos Wigg bought a deposit, estimated to contain 10,000,000 tons, for $12,500. Trajano de Medeiros bought another estimated to contain 12,000,000 tons for $12,500. E. Thum bought the Casa de Pedra deposit, 2,000,000 tons, for $15,000. This is probably an incomplete list, and the statements made in the consular report may not have been exactly correct (since it is difficult to obtain such information) but it will serve to illustrate the effect produced by the publication of Derby's paper.

The Itabira Iron Ore Co. group already controlled the railroad which, leaving the port of Victoria, crosses over into the valley of the Doce river and follows it to Serra Escura, about 275 miles. A project was gradually evolved to extend this road about seventy-five miles to Itabira, construct ore-handling equipment at the port, and to provide a fleet of vessels to take 3,000,000 tons of ore to Europe and 2,000,000 tons to the United States yearly, bringing back coal to Brazil, the idea being that the main output of ore would go to Europe and the United States for smelting and steel manufacture, but sufficient coal would be brought to Brazil to permit smelting enough ore there to supply as much of a market as could be developed in Brazil and elsewhere. It is estimated that ore containing 68 per cent iron can be delivered in Europe and the United States for $6.80 per ton, which would permit its use in competition with existing ore supplies in those regions.

But various difficulties appeared in the execution of such a project. It developed that the railroad could not simply be extended, but would have to be rebuilt throughout in order to permit economical ore-handling, and the cost of doing that would be $40,000,000 to $50,000,000. The port of Victoria is in the state of Espirito Santo while the ore deposits are in the state of Minas Geraes, so the ore in moving to the port would have to cross a state boundary. Brazil, like the United States, is a federated republic of self-governing commonwealths. Its Constitution contains a section almost identical in wording with that section of our Constitution which has been construed by our courts as forbidding any imposition of duties on the interstate movement of goods. It was not at first so construed in Brazil and duties were until recently collected on interstate shipments. In the case of shipments of iron ore from Minas Geraes to São Paulo, an export tariff of seventy-five cents per ton was at first demanded; even if a reasonable rate could be obtained the right of the state to increase it constituted a discouraging factor from the business standpoint until, by presidential decree forbidding interstate taxation, it was eliminated in May, 1932. The experience of the manganese producers with freight rates and port charges was also not encouraging,[18] nor was the necessity of dealing with two state governments as well as with the national government. It was reported that the latter was disposed to require the enterprise to construct a larger iron and steel plant in Brazil than seemed to be justified by the markets available. Since the capital for carrying out the project would have to be secured abroad it is more difficult to resolve

[18] See p. 152.

these political economic problems than if domestic capital was involved. At any rate, by 1934 none of the groups referred to in the preceding paragraphs had yet embarked upon large-scale development of their iron-ore holdings.

Aside from increase in the production of pig iron by the charcoal furnaces in Minas Geraes later described, the most interesting iron and steel production in Brazil after 1910 was that of the Companhia Electro-Metallurgica Braziliera, organized at Riberão Preto, São Paulo, in 1917. The Companhia de Electricidade at that place was organized to supply electric light and power to the town and the neighboring coffee plantations. It soon found that power demand in the town was negligible, while that of the coffee plantations was limited to a small part of the year; the demand for light was, as always, limited to a few hours daily. The power company conceived the idea of establishing an electrometallurgical plant that would provide a large steady demand for power, and thus give it a better load factor. With the aid of the government the company referred to was organized, with the power company as its principal stockholder.

Two twenty-five ton electric pig-iron furnaces of Swedish type were constructed, two six-ton electric converters to convert the pig iron into steel, a six-ton steel refining furnace and other accessory equipment were provided. The iron ore was derived from Morro do Ferro, in Minas Geraes, about 250 miles distant by railroad, and involving a transfer from narrow to broad gauge. At first Minas Geraes proposed to place an export tax of seventy-five cents per ton on the ore, but because of government participation in the enterprise this was reduced to a nominal figure. The electric resistance furnaces were able to furnish the heat requirement of the metallurgical operations, but the carbon necessary to reduce the iron oxide of the ore was "obtained with some difficulty and at high prices from the neighboring country."[19] Not only was the supply of wood short and the nature of the wood unfavorable for charcoal-making, but the wide range of size of the pieces made it difficult to produce charcoal of as uniform a quality as desired. The ore, obtained from the top of the deposit, was also somewhat variable in quality and working on so small a scale makes it difficult to obtain uniformity through mixing, as can be done in large-scale operations. It was not only somewhat difficult to produce steel of high quality, but there was no local market for it, the demand being rather for pig iron, and cheap mild steel. The enterprise was therefore in somewhat the position of a transportation enterprise employing Rolls-Royce automobiles for a service that could be adequately performed by buses; in other words, an expensive[20] plant with small capacity and relatively high operating costs reached for a market that could be served by plants of large capacity and low operating cost. A report on it made in 1929 said it was producing steel by remelting scrap iron, and there are no definite figures as to how much pig iron it produced from ore. It has since definitely gone into liquidation and, after standing idle for some years, was taken over by the São Paulo state government.

Some recent reports to the effect that about 70,000 tons of pig iron are now being produced in Brazil yearly are undoubtedly erroneous. The charcoal furnaces that produce pig iron in Minas Geraes mostly ship it to found-

[19] Teixeira, *loc. cit.*, p. 733.

[20] By the time it began operation in 1922 about $1,000,000 had been expended on it.

ries in São Paulo and Rio de Janeiro for the production of iron castings. The Brazilian phrases for pig iron and cast iron are so similar that a ton of iron is counted twice, once as pig iron and again as cast iron. The actual output of pig iron probably does not exceed 30,000 tons yearly, from twenty- to twenty-five-ton charcoal furnaces in Minas Geraes. The most important of these is the Companhia Siderurgia Belgo-Mineira, whose plant at Sabará not only makes pig iron but also makes steel in a small open-hearth furnace; a capacity of 36,000 tons per year is claimed for it. The Companhia Querroz, Jr., Limitada, with plants at Esperanca and Burnier, claims a capacity of 18,000 tons of pig iron yearly. The Companhia Brasileira de Metallurgica has a charcoal furnace at Morro Grande and a plant at Nictheroy, across the bay from Rio de Janeiro, in which steel and castings are made; a yearly capacity of 12,000 tons is claimed. The Companhia Ferro Brasileira, at Gorceix, claims a capacity of 12,000 tons of pig iron; the Companhia Metallurgica Santo Antonio, at Rio Acima, 7,000 tons; the Companhia Mineração e Metallurgica S. Caetano, at São Caetano, 10,000 tons; and the Metallurgica Magnavaca 7,000 tons. The actual combined pig-iron production of all these plants is about 30,000 tons yearly, or the quantity that a single modern blast furnace in the United States would produce in a month. The dilemma is that the charcoal supply is neither large enough nor good enough to permit any major development of charcoal iron-making, and no coke is available for ordinary large-scale pig-iron production. A large-scale project like that of the Itabira Iron Co. cannot develop out of embryonic operations by slow natural growth; it must be set up on a big scale to be feasible at all, and that involves investing something like $100,000,000 before any returns can be hoped for. The anticipated operating profit of such an enterprise is only of such an order of magnitude that it could be wiped out by increases in local taxation, export tariffs, and other overhead charges that would not be under the control of the company. The political aspects of the enterprise are therefore as important as the natural conditions that govern it.

The study given to the iron deposits of Brazil since the publication of Derby's paper indicates that his estimates were too small rather than too large. Some workers have made estimates double his; the estimate of the Brazilian Geological Survey is 8 billion tons, and conservative estimates give it at least 4 billion. The deposits occupy a region of 6,000 square miles and are probably the largest and richest of the world's iron-ore reserves. Nothing has been said of other iron-ore occurrences in the states of Bahia, Espirito Santo, Matto Grosso, Goyaz, Paraná, Santa Catharina, and Rio Grande do Sul, for which claims are made. But even if they should prove on thorough exploration to be no more promising than those of São Paulo, referred to above, Brazil would still be one of the world's greatest storehouses of iron ore. It seems clear that no considerable domestic industry can ever be developed on the basis of using charcoal as fuel or by using electric energy for heat and charcoal as the reduction agent. It is by no means certain that any large industry can even be developed on the basis of imported coking coal, exporting iron ore as return cargo for the ships that bring the coal. Merely to export the ore for reduction abroad is not a popular national policy in Brazil, and what the final outcome of discussions will be it is unsafe to attempt to prophesy. Iron production in Brazil now is probably not more than a quarter of the domestic requirements, and until those are met it would be precarious to forecast whether Brazil can become an iron and steel exporting country. It is difficult, in addition, to be certain exactly what its requirements amount to. The reported figures of iron and steel imports (until recently about 100,000 tons yearly) are merely those of commodities so classed, and manufactures of iron and steel, valued at $34,472,000 in 1929, are separately listed. Quite possibly manufactured articles would continue to be imported, even if domestic raw materials of manufacture were available in larger amount and variety, but, on the other hand, domestic manufacture would be encouraged if raw materials of good quality were available at a lower price than imported materials. Whether they can be locally produced at a lower cost than delivered imports is, of course, the crux of the problem. Experience so far seems to indicate that they cannot, but if transportation facilities for the export of ore are provided the cost of imported fuel may be lowered sufficiently to permit its economical use.

MANGANESE

Manganese is a mineral that comes to the attention of the general public chiefly through tariff debates, because it is not employed in ways that make it visible. Its chief use is in connection with steel-making, where it is mainly employed in a way that, like the use of soap in washing one's hands, does not make it appear in the final result. In addition, it is used in making manganese steel and for a great variety of metallurgical and chemical uses. The tonnage used in the world yearly is so large as to make it one of the major mineral substances; of the ore 95 per cent is used in connection with the steel industry. It occurs widely throughout the world and commercial conditions are such that only ore that contains about 50 per cent of the metal (or can be concentrated to that richness) has any considerable market. It is a curious natural fact that the regions in which such deposits occur are all remote from the principal centers of consumption; regions of steel and chemical manufacture. Brazil is one of the four principal regions of the world in which large deposits of manganese ore occur. It is true that they are in close proximity to the iron-ore deposits of Brazil, so that if and when a steel-making industry develops there a local market will exist for them, but at present there is none, as has been explained above.

Nothing was known of the existence of this ore in Brazil until 1888, when a deposit was found near Burnier in making a railroad cut for the Central of Brazil railway. Carlos Wigg secured some lands there and began producing the ore for export in 1893. Within the next six years two more operations started nearby, and in 1900 mining began on another deposit discovered near Lafayette, eighteen miles distant. Many additional discoveries have since been made in this general region, which has produced over 90 per cent of the manganese-ore output of Brazil to date. About the same time ore was discovered near Nazareth, Bahia, and there has been a smaller production from that region. A deposit is known to exist in Matto Grosso, but is too remote to operate in competition with the others.

In 1913, the world was drawing the greater part of its manganese-ore supply from Russia, where excellent deposits occur in Georgia. The next largest source of supply was India, where there are numerous deposits of good quality. The outputs for 1913 were: Russia, 1,310,064 metric tons; India, 828,128 tons; and Brazil, 122,300 tons. The ore usually sold at the port of final delivery at $10 per ton of 50 per cent manganese material, the actual price depending on its manganese content, and a penalty being imposed if it exceeded 8 or 9 per cent silica, and 0.15 per cent phosphorus. Although many of the Indian deposits are about twice as far from the seaport as those in Brazil, the freight rate was the same; the Russian deposits were only about one-third as far from seaport as the Brazilian and the freight rate was correspondingly lower.

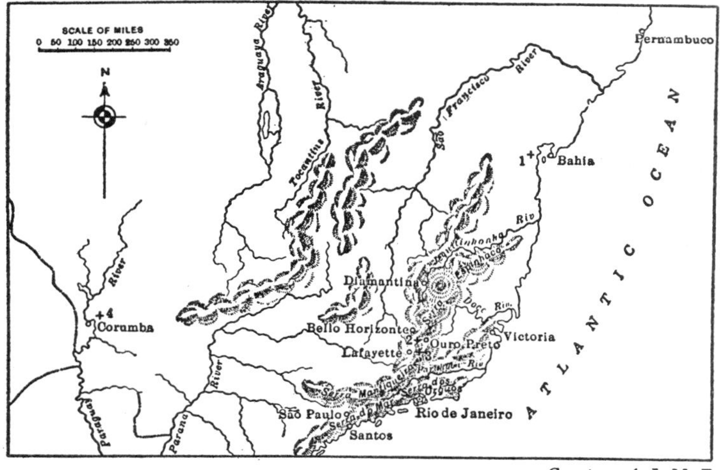

Courtesy A. I. M. E.

FIG. 4. MANGANESE DEPOSITS OF BRAZIL

Up to the outbreak of the World War in 1914 an abundant supply of manganese ore of acceptable grade was available, the districts competing in the world markets on the basis of quality and delivered price. Soon after the outbreak of the war, and especially after our entrance into it in 1917, the situation materially changed. The shortage of shipping, and the German submarine campaign which caused it, discouraged exports from Russia and India and stimulated exports from Brazil. In

1916, the production in Brazil was 495,172 long tons, and in 1918 the United States alone imported 351,428 metric tons from Brazil, or about double its prewar production. The producers who had been operating increased their output, and others entered the business. Most of them were Brazilians, but some foreign capital was involved, and the United States Steel Corporation eventually bought the Companhia de Mineracão at Lafayette, from a Brazilian company, Cia Morra da Mina. The total production in Brazil from the beginning to 1921 was about 6,000,000 tons. After the close of the war this situation favoring Brazil no longer existed. The price of ore had soared to unprecedented figures during the war, furnishing a corresponding incentive both for the old producing regions to regain their former position, and for the development of new deposits elsewhere. The most important of the latter was in the Gold Coast Colony, West Africa, where good ore was discovered within thirty-five miles of the seaport; exports from there reached 344,800 tons by 1926. Exports from Russia in 1926 were only about half what they were in 1913 and those from India about three-quarters. Egypt also entered the ranks of manganese producers and by 1929 the relative outputs of the different countries were as follows:

	Metric Tons
Russia	1,200,000
India	1,010,237
Gold Coast	465,282
Brazil	316,172
Egypt	191,477
All others	276,832
	3,460,000

That year was the high-water mark in world production of manganese ore. The depression which began near its close affected manganese consumption through curtailing steel output. The exports in 1930 were only about two-thirds what they were in 1929 and Brazil's output was correspondingly affected. The Russian deposits, being owned and operated by the Government, continued to produce for export in the face of falling prices, a form of competition which private producers in the other countries were unable to meet. Russia's output in 1930 was nearly one-quarter larger than in 1929.

The result of this was that by the middle of 1932 all the manganese mines in Brazil had ceased producing. When and if they will be able to resume production will depend, of course, on how soon steel production comes back to normal and how keen the competition of other producing regions is during the period of revival.

The manganese deposits about Lafayette are large lenticular bodies of manganese carbonate that have been altered to oxides by weathering to a depth of as much as 400 feet. In the Burnier district the oxides are interbedded with iron-bearing sediments. The grade of ore as shipped in Brazil ranges from 38 to 50 per cent manganese and some other countries are able to maintain a closer approximation to 50 per cent. The Brazilian ore tends to be higher in silica and iron than some of the others. The only possible way for the output in Minas Geraes to reach seaboard is over the government-owned Central of Brazil railway, nearly 300 miles, to Rio de Janeiro. Until recently, at least, the government seems to have seen little necessity to facilitate production for export.

In 1926, the export tariff on manganese ore in Brazil was $2.70 and the state tax added to this made a total of about $5 per ton export tax. As against this the export tax on the Gold Coast was about fifty cents per ton, and the government royalty in India (no export tax) was about six to ten cents per ton. Freight rates in Brazil were about the same as for twice the distance in India, where improvements by the government of the port facilities at Vizagapatam will cut down railroad transportation to a very short distance after 1933. Port charges in Rio de Janeiro were relatively high and a minor but illuminating detail is that the freight charges on the Central of Brazil railway are made up of four different items, including a charge for collecting the charges. Probably the temporary cessation of manganese production in Brazil will be effective in bringing about some change in this general attitude of "charging all the traffic will bear." It is politically difficult to revise government charges downward when changes in world conditions make traffic less able to bear charges than it once was. The Minas Geraes export tax was later reduced to 10 per cent of the assessed value of the ore, and presumably will be eliminated by the national decree of May, 1932. On the other hand, the Federal government made an attempt to increase its export tax in 1930.

Another significant factor is the increase in world capacity for production. The entrance of the Gold Coast and Egypt into the list of major producers has already been mentioned. The wave of nationalism that has swept the world since the Great War has caused all the principal consuming countries to seek sources of manganese-ore supply in regions that are at least under their political control. A group of the principal consumers in England now control about two-thirds of the productive capacity in India, while a recently discovered deposit in the Union of South Africa promises to develop into another large producing region, nearly 150,000 tons having been produced in 1930. French consumers are actively developing productive capacity in the regions in northern Africa that are under French control, while some German consumers are seeking to develop production under their control in southern Asia Minor.

In the United States, which is the largest importer of manganese ore, domestic interests have succeeded in securing an import tariff that at the time of writing amounts to 100 per cent of the cost of foreign ore delivered at Atlantic ports. So far this has not resulted in developing a domestic output that amounts to more than 10 per cent of our annual requirements, but domestic interests insist that it is merely a matter of time and the continuance of tariff production until a substantial domestic production will result. Competition in world markets is thus likely to be increasingly difficult for Brazilian producers, and the question of ocean freight rates becomes of critical significance. These not only ordinarily vary over a wide range from year to year, but where shipping lines to a producing region are under the control of the same government that wishes to stimulate production in that region it is not difficult to secure advantageous rates. It seems clear, therefore, that development of manganese production in Brazil, in the near future at least, is likely to be contingent on recognition there that it is a keenly competitive business which will require a similar solicitude on the part of the government as manganese production is receiving in the other producing regions of the world.

Construction Minerals

The commodities already discussed exhaust the list of minerals of which there has been any considerable production in Brazil, except those non-metallic minerals for construction purposes that normally occupy a large place in the world supply of mineral substances, though they are commonly almost ignored in discussion of mineral problems. Stone, gravel and sand for construction and clays suitable for brick-making and ceramic industry are so widely distributed throughout the world that they are typically minerals of relatively local production and only for a few countries of the world are statistical data available as to the quantities and values involved. Brazil is not one of them, and reference to them can be couched only in the most general terms. Broadly speaking, the natural resources of such minerals in Brazil seem adequate to the demands that have developed, or are likely to develop there. It should be noted, though, that brickmaking and ceramic manufacture require the use of fuel, and its general shortage in Brazil is something of a handicap to those industries. This is especially true of Portland cement, which has a heavy requirement for fuel for its burning and the mechanical operations incidental to its manufacture. As a result, cement production has only recently developed in Brazil; the importance of imports is well shown by the following table:

Cement Imports, Brazil

	Barrels	Value
1927	2,591,000	$6,982,000
1928	2,675,000	6,843,000
1929	3,139,000	7,400,000
1930	2,290,000	5,058,000
1931	670,000	1,276,000
1932	941,000	1,293,000

Three cement plants have been constructed, one at Perús, in São Paulo, another at Juiz de Fora in Minas Geraes, and a third at Nictheroy, in Rio de Janeiro. The Perús plant has a capacity of 1,000,000 barrels yearly but has only produced half that much. Whether these plants represent a real economic advance depends on whether production costs in Brazil plus the cost of distribution to consumers are less than delivered cost of imported cement, assuming that the local plants are able to produce a product of equal quality with the imported. The problem then becomes largely one of maintaining steady operation, since production costs are seriously affected by irregularity of operations in any industry where the investment cost is high per ton of output. Cement plants ordinarily control their own fuel supply and a plant which has to buy fuel in the open market incurs an additional factor of uncertainty in production cost.

Copper, Lead and Zinc

These three metals are major items in world mineral production, since although their production is ordinarily reported in terms of metal produced they are typically obtained from low-grade ores that have to be concentrated, consequently the amount of mining that has to be done is much greater than might be easily realized from the metal production figures. The approximately 2 million tons of copper metal produced in 1929 was the final product of mining some 100,000,000 tons of ore.

No important production of any one of these three metals has ever developed in Brazil, although ore occurrences have been reported from many places. Copper is said to have been produced in Bahia as early as 1783; there has been a small production at intervals since but none recently reported. In 1900, a Belgian syndicate, Société de Mines de Cuivre de Camaquam, opened a mine and constructed a power plant and smelter on the Camaquam river, in Rio Grande do Sul, but the enterprise ended in failure, as did a subsequent attempt to ship concentrated ore to England for smelting. Lead is claimed to be of frequent occurrence in Silurian limestones, but the only place at which any attempt has been made to mine it is at Riberão da Prata, near Blumenau, in Santa

Catharina. Large claims were made for this deposit about 1914, but an attempt to interest German capital came to nothing. Lead-ore production in recent years is reported as follows: 1924, 252 tons; 1925, 251 tons; 1926, 320 tons; 1927, 787 tons; 1928, 463 tons. No discoveries have been made of zinc ore that have even been claimed to be workable.

Precious and Semiprecious Stones

Diamonds have been adequately discussed in the opening sections of this chapter, but something may be said about the other varieties. And since the dividing line between precious and semiprecious stones is somewhat uncertain, both groups will be considered together.

A single mineral substance, beryllium aluminum silicate, is variously known as emerald, aquamarine, and beryl, according to its color. The green variety, emerald, is seldom free enough from flaws to be used as a gem stone, and while some of it is found at various places in Brazil it is not of economic importance. But the sea-water blue variety, called aquamarine, and the yellow or brownish beryls occur there in considerable amount; many places in Minas Geraes are famous for their aquamarines, the finest ones known having come from there. They are produced by small-scale operations, and the cutting of them for market is done by small shops in Bello Horizonte and Rio de Janeiro, that both cut stones and work them up into jewelry; quantities of the material as mined are also exported for similar treatment abroad.

Beryl is the only commercial source of the metal beryllium, which has recently attracted a good deal of attention because of possible applications in modern metallurgy. As the pure mineral contains only about 4 or 5 per cent beryllium, and production of the metal is now somewhat difficult, it is quoted at $25 to $50 per pound, but if a demand develops the price of the metal can undoubtedly be decreased, and a greater demand created for beryl. Large crystals not of gem quality are found in Brazil; one was seven feet long and weighed three tons. L. J. Moraes has recently described the Brazilian deposits in some detail.[21] If the various technical problems connected with the metallurgical use of beryllium can be successfully solved, its production in Brazil may greatly increase in importance.

Brazil is a famous place also for the occurrence of topaz, again in Minas Geraes, though the stone is found in other states also. Tourmalines are found in the same general regions. Amethysts of good quality are found in Rio Grande do Sul, Bahia, and Minas Geraes. Garnets of gem quality have been found in various states. The greatest amount of business, by volume, is in agate, which is picked up as pebbles along the rivers in Rio Grande do Sul and Ceara, and exported for coloring and polishing, and in quartz crystals. The production of semi-precious stones, the world over, is typically neither a large nor very profitable business, and beyond recognizing its existence little needs to be said about it.

Recent figures as to precious-stone production in Brazil are as follows:

	1924	1925	1926	1927	1928	1929
Diamonds, industrial	$677,670	$983,174	$1,088,528	$1,205,178	$1,310,343	$815,816
Diamonds, gems	630,328	417,138	755,752	405,110	493,152	269,641
Other gems . . .	127,989	59,983	43,226	37,131	68,596	27,655
Agate, metric tons	154	77	130	130	161	483
Crystals, tons	203	150	161	269	309	499

[21] *Economic Geology*, vol. 28, pp. 289–292, 1933.

Other Minerals

There are various miscellaneous mineral substances that have come to be of technical and social importance in the past century, of which the world does not require any great amount, but of which the resources are worth listing. So far as Brazil is concerned they all would be commodities for export, since little or no domestic demand exists for them. Those of which workable deposits exist add to the exportable surplus of the country, and are correspondingly important there.

Arsenic. Apparently all the white arsenic produced in Brazil is as a by-product at the St. John del Rey gold mine, although the Ouro Preto mine at one time also produced some. The St. John del Rey yielded 148 tons in 1924 and normally produces about 200 tons yearly. Exports have been reported as follows: 1924, 240 tons; 1925, 69 tons; 1926–1928, none.

Asbestos. About a third of a ton of asbestos was reported as produced in 1924 and 1925, but none since.

Graphite. In 1929, an output of sixteen tons was reported; previous reports had been negligible.

Chromite. Chromite was not much in demand before the World War, although its use had been slowly growing for a century. The United States, Germany, France and England were the principal consumers, since it finds its main application in connection with the iron and steel industry. The then supply was mostly obtained from New Caledonia, Rhodesia, and Russia. As in the case of manganese, the war provided an incentive for production in Brazil, and in 1918 some 13,500 metric tons of it were produced in Bahia, in the Bomfin district, where it occurs in the Archean rocks. Search for deposits throughout the world was so stimulated by war conditions as to result in considerable additions to the known reserves, which promise an adequate supply for many years to come. When normal competitive conditions were resumed after the war the Brazilian deposits were mostly unable to compete with the others where the deposits were larger and facilities for production and transportation were more favorable. Recent production in Brazil has been reported as 1,500 tons in 1926, 1,820 tons in 1927, twenty tons in 1928, and seventy tons in 1929. L. C. Ferraz gives a list of all the places where its occurrence has been noted in his Compendio dos Mineraes do Brasil (Rio de Janeiro, 1928).

Monazite. Monazite is a good example of a mineral which attains commercial importance and then loses it again. About 1885, Welsbach, of Vienna, made a great improvement in gas lighting by saturating a cloth "mantle" with a solution of salts of the elements known as the alkaline earths. After the cloth had been burned away the alkaline earth oxides retained the shape and emitted a brilliant white light instead of the ordinary yellow flame. About that time John Gordon discovered heavy sands along the sea beach south of Belmonte, Bahia, which were identified as the mineral monazite, a phosphate of cerium, lanthanum, didymium, and thorium. For a time sea-beach deposits there and elsewhere along the coast of Brazil were an important source of world supply of the alkaline earth elements. But the development of the metallic-filament electric light soon made gas lighting obsolete, and the principal demand for these elements went with it. Some other minor uses developed (the "flint" in a cigar-lighter is a cerium-iron alloy) but the present world demand for these elements is quite small, and the latest reports from Brazil indicate that production of the raw material has ceased.

Zirconium. Zirconium is an element for which various uses have been found in recent years; the mineral zircon, in which it occurs, was originally used as a semiprecious stone. None of its uses yet requires large amounts and though there are no accurate figures of world production they certainly are less than 5,000 tons of zirconium mineral yearly. India and Brazil are the principal producing regions, and the following figures are given for the production of the latter:

ZIRCONIUM SAND PRODUCTION, BRAZIL

	Tons	Value
1920	597	$25,034
1921	134	6,095
1922	366	16,844
1923	36	883
1924	842	26,137
1925	137	4,142
1926	12	563
1927	285	8,313
1928	913	31,048
1929	1,187	41,419
1930*	237	
1931*	137	
1932*	815	

* Exports.

Production of zircon in Brazil seems to have had its origin as a by-product in the production of monazite, and declined with it. Another zirconium mineral, baddeleyite, has been found in Brazil in considerable amount, and has been mined in the Caldas region, on the border between Minas Geraes and São Paulo. Zirconium minerals occur at many places in Brazil and if a considerable world demand develops for them their production might become important.

Ilmenite. This is a mineral for which a demand has recently developed for making titanium oxide (white pigment). In 1929, two companies, the Soc. Min. et Industrielles Franco-Bresilienne, and Mancio Israelson attempted its recovery from beach sand in Esperito Santo and John Gordon similarly produced it in Bahia. Reported production is as follows:

Ilmenite, Tons

1924	210
1925	1,500
1926	1,498
1927	1,307
1928	2,102
1929	6,361

Nickel. Nickel deposits of workable grade and large size are of rather infrequent occurrence, and the immense deposits in Canada have something of a natural monopoly of world production. Nevertheless, there is a small nickel production in various other parts of the world, among them being Livramento, Minas Geraes, where the Companhia Nickel do Brasil has for the past few years been mining a deposit of garnierite, which occurs associated with serpentine and peridotite rocks. Reports from Brazil state that between 1919 and 1923 the average annual output was 15,000 pounds of nickel and since 1927 has been as follows:

Nickel Production, Brazil

	Pounds	Value
1927	16,000	$ 5,100
1928	40,000	18,000
1929	31,700	15,000
1930	114,400	48,000
1931	7,000	

Since the demand for nickel in Brazil cannot be large this relatively small output must, in part at least, be sold in world trade in competition with the principal producers.

Mica. Mica claimed to be of excellent quality is found at various places in Minas Geraes, and elsewhere in Brazil, and for some time there has been a small production. That in 1916 was reported as about 110,000 pounds, valued at $55,859. "International Trade in Mica" (Trade Promotion Series, No. 95, United States Dept. of Commerce), pp. 23–24, describes in detail the situation there in 1929 when there were five producing mines. Exports in 1924 were valued at $137,810, but by 1928 had declined to $53,177. In 1932, exports were 92,593 pounds as compared to 97,000 pounds in 1928. It is of interest chiefly because it is prepared for market in Brazil, instead of being exported in the raw state.

Bauxite. Bauxite deposits are important in British and Dutch Guiana, where the general geology is similar to that of eastern Brazil, but much less is known about their occurrence in Brazil. A few paragraphs concerning it appear in Feraz's "Compendio dos Mineraes do Brasil," already cited. Among the localities he lists are four in Minas Geraes, one of which, Motuca, about fifteen miles from Morro Velho, is owned by the St. John del Rey company. Other occurrences are in Amazonas, Pará, and Parahyba do Norte. An interesting deposit is known to exist in Maranhão, where bauxite has apparently been impregnated with phosphate material that has converted it into aluminum phosphate. As the world demand for bauxite increases it is quite possible that more intensive study of the Brazilian occurrences will lead to their commercial exploitation.

This by no means exhausts the list of mineral substances that occur in Brazil, but all those of which the present or prospective production seems of any importance have been touched upon. Those interested in a more complete list of occurrences should consult technical literature, especially the chapter on Brazil in Miller and Singewald's "Mineral Deposits of South America," which concludes with a bibliography complete to 1918, and also L. C. Ferraz's "Compendio dos Mineraes do Brasil" (Rio de Janeiro, 1928).

Conclusion

The fundamental weakness in the mineral situation in Brazil is the practical absence of commercially exploitable deposits of mineral fuels. Imported fuel can be delivered at seacoast cities at a not unreasonable price, but the cost of transportation inland is a serious handicap to the development of any enterprise which requires fuel, such as the iron and cement industries. The handicap of high fuel cost operates against many industries. Not only metal fabricating but general manufacturing operations feel its effects. The development of industrial civilization in Brazil will have to work within the limits thus set.

Among the handicaps to which mineral developments have been subjected property titles must be mentioned. Large tracts of land have descended by inheritance and in some cases have been informally divided by the heirs without any formal partition proceedings. Under the Brazilian law an heir to an undivided estate can enter the property, and this is capable of giving rise to many difficulties when a tract is purchased from its apparent owners. The difficulty is not simply a theoretical one, as purchasers of tracts have sometimes had the sale canceled, losing the improvements they have put on it.

Until recently the state import tax, through the ease with which the amount could be varied, has been a sort of sword of Damocles over the heads of mineral producers. If a venture in mineral production proved successful there was no assurance that the export tax would not be so increased the following year as to wipe out the margin of profit. The recent decree abolishing interstate taxes will presumably relieve the situation to that extent, but since many minerals must be exported to find a market the danger of a Federal export tax still exists.

Transportation costs are a handicap to mineral production also. Both ordinary roads and railroads are few in proportion to area. On the railroads, freight charges are not only relatively high, but are computed on a theoretical valuation of the shipment that is so fixed as to make low-grade ores take an unduly high rate. These factors, of course, are under social control, and might be altered under a policy of promoting mineral development.

Labor conditions have been reviewed in the section discussing gold mining, and the points there made need not be repeated. In industrial countries the modern trend is to perform labor by machinery, using men only for work that is essentially supervisory in character. The common laborer of Brazil is not at present well adapted to the latter method; how he will react to future development no one can certainly say. All the evidence seems to indicate but little hope of Brazil's ever becoming a country in which there is large and important industrial development based on domestic mineral production. The mineral-fuel shortage is a general handicap, labor conditions are not favorable, and in the case of many minerals the basic geological conditions are also unfavorable. The present world-wide general trend is toward national industrial independence, but Brazil's natural situation is such that it must look toward international exchange as the policy best suited to promote its interests for some time in the future.

Manganese ore is the only mineral commodity listed separately among the exports of Brazil. Though ranging from 189,088 to 356,116 tons in the period of 1927–1930, it corresponded only to from one-third of 1 per cent, to 1 per cent of total exports by value. Coffee corresponds to two-thirds or three-quarters of total exports and the rest are all agricultural products. Exports of manganese ore in 1929 were valued at $3,375,000; imports of coal, petroleum products, unmanufactured iron and steel, and cement were valued at over $60,000,000. In 1932, they were $93,000 and $17,263,000 respectively. Brazil is, therefore, not an important mineral-producing country—its total annual mineral production, both by tonnage and value, is less than that of our state of Florida, not ordinarily thought of as a mineral region. Except for iron and manganese, Brazil is likely to remain permanently an agricultural rather than a mineral region.

CHAPTER VII

URUGUAY AND PARAGUAY

URUGUAY'S 72,172-square-mile area makes it the smallest of South American countries, but its nearly 2 million of population, 90 per cent white, its large proportion of utilizable land, and its transportation facilities which, in proportion to area, are the best in South America, make it a region of high activity and wealth.

Geologically, it represents the southern tip of the ancient highland region of Brazil, here so low-lying that only in the north does it reach an elevation of as much as 1,000 feet, even though it is in places overlain by Tertiary deposits. The north is a region of rolling grazing land; to the south and southeast Quaternary and alluvial deposits make it a fertile agricultural land. With short mild winters and summer temperatures that are not excessive, it has a climate that is between temperate and tropical. Between 1888 and 1930, its population increased three times.

Its total exports in 1929 were valued at $91,486,000, but in 1931 declined to $27,420,000. Two-thirds of 1929 exports were animals and animal products, and practically all the remainder agricultural products. Curiously enough, its only mineral exports are sand and stone, two substances which because of their low unit value are seldom important factors in international trade. But neither sand nor stone are available in Buenos Aires and the adjacent populous region, and the demand for them there is supplied mostly by Uruguay. Exports of sand in 1929 reached 2,262,113 tons, valued at $2,320,000 and of stone 498,031 tons, valued at $694,000, but declined to $906,000 and $424,000 in 1932. As these figures are not much smaller than the official figures of total production, nearly all the output must be exported.

Aside from a few thousand tons of clay and about a thousand tons of talc yearly, there is no other mineral production in Uruguay worth mentioning. No gold has been reported as produced since 1926, and before that ranged from 75 to 150 ounces yearly. There is no coal, petroleum, iron, copper, lead or zinc production, the only other item appearing in the official statistics being 40 to 100 tons of agate yearly, usually shipped to Germany for cutting. Miller and Singewald list more than a dozen publications on the mineral resources of Uruguay and the government maintains an Instituto de Geologia y Perforaciones. This was established in 1912, but the figures cited above are sufficient indication that even government concern for the development of domestic mineral resources can do but little where the natural conditions are naturally unfavorable.

In the northern provinces, particularly Rivera, gold and copper have been mined in a small way at various times and iron and manganese are reported to be present. In Minas, one of the southern provinces, still other iron-ore bodies are present but in the light of present knowledge and economics none of these seems likely to prove important. Rolf Marstrander, who is familiar with the country, has succinctly reviewed the geology and mineral resources[1] but stated truly that "A real mining industry never has existed in Uruguay nor does it exist now." At

[1] *Mining Magazine;* London, June, 1916.

the time he wrote only one metal mine and one talc mine were in operation.

In 1929, Uruguay imported 371,031 tons of coal, valued at $4,672,000 and of fuel oil 2,139,000 barrels, valued at $2,930,000. Kerosene imports were valued at $1,926,000, gasoline $9,643,000, gas oil $372,000, lubricating oil and grease $654,000, asphalt $413,000, iron and steel $6,400,000, and Portland cement $484,000. There is a small domestic Portland cement plant, using fuel oil, and the Ancap, a subsidiary of the Instituto Geologia y Perforaciones, has made contracts with German technicians for exploratory drilling in the northeastern corner, where it is hoped to find petroleum in Permian black shales, or else an extension of the beds which in Santa Catherina, Brazil, contain coal. Exploratory drilling is also being done on the western border, near Paysandú. While it is possible that this work will result in development, the outlook is discouraging. A more promising project of the Ancap is the construction of a domestic petroleum refinery which is planned to operate on crude petroleum brought from Peru. Bids for its construction were called for in 1932 but construction has not been started. The potential water-power resources of Uruguay have been estimated at 300,000 horse-power but none has yet been developed, though plans have been made for a site on the Rio Negro, 150 miles from Montevideo. As in Argentina, most of the imported coal comes from Great Britain for the same reasons as set forth on p. 177. It is the third largest coal-importing country in South America.

While the efforts of the government to promote domestic production of minerals for domestic consumption will probably have some beneficial effect, nothing can greatly affect the natural condition that Uruguay, so far as known, does not possess workable deposits of the principal minerals. The industrialization of the country must principally be based upon imported fuel and structural minerals, to be obtained through the export of animal and vegetable products.

Paraguay's area is variously estimated at from 97,700 to 176,000 square miles, since the boundary between it and Bolivia has always been disputed and in 1928 led to war between the two countries. Roughly, Paraguay is about three times the size of our New England states and is a region of vast forests and plains, low-lying and with a sub-tropical climate. Whether it is really so poor in mineral resources as published information would seem to indicate is perhaps open to some doubt. It has less than a million inhabitants, of whom 90 per cent are mestizo or Indian. It has no internal means of transportation except its rivers and about 500 miles of railway, mostly feeders to the ports on the rivers. Of its total exports, valued at $11,839,000 in 1930 and $7,523,000 in 1932, quebracho extract, canned meat, hides, tobacco, and yerba mate constitute two-thirds and the other third is wood and a variety of agricultural and pastoral products.

The economy of the country, a simple agricultural and pastoral one, has little need of minerals and, hampered by lack of transportation, no mineral deposits of a size or quality that would permit production for export have been found, nor is there much incentive to explore for them. There is some iron ore and in 1854 the government built a one-ton charcoal furnace at Ibicuy as a beginning of a domestic iron industry. The reasons why it does not flourish will be evident from what has elsewhere been said about iron production. Copper, zinc, and mercury ore occurrences have been mentioned, but there are no reported occurrences of gold, silver, or lead. Various non-metallic minerals are known to occur and there is an abundance of building stone. Sulphur is produced on a small scale. There are no official reports of mineral production nor any official estimates. There are no detailed figures as to imports, but apparently small amounts of coal and petroleum products are brought in, the latter under an import duty. The railroads use wood for fuel, and while there is considerable potential water power the sites are mostly in unsettled regions. Of all South American countries Paraguay is the one in which minerals have been of the least importance, and it seems probable that it will long occupy that position.

CHAPTER VIII

ARGENTINA

ARGENTINA is best known for its pampas, those great all but flat plains sloping gently up from the Paraná and the Atlantic to the foot of the high Andes on the west. When first seen by white men the pampas were treeless, grass-covered and seemed to stretch away without end. Now much of the area has been brought under cultivation, for the region is one of rich soil, adequate rainfall, nearness to the sea and interposes no topographical difficulties in the way of railway building. It has become one of the world's great sources of grain and meat. It is this part of Argentina which is seen by most visitors to the country as they travel by plane or train the 550 miles from Buenos Aires, the great port, to Mendoza. It is a journey not unlike that from Kansas City to Denver, passing similarly from well-cultivated farming land across progressively drier almost desert areas and finally into the fertile valley between the foothills and the main mountain range.

South of this middle belt is a less favored zone of smaller rainfall. The rivers are few and the grass is shorter. This is Patagonia, and it, in many particulars, resembles the larger part of Wyoming. Like the latter it is extensively devoted to sheep raising. North of the pampas there are three Argentinas, and at the west a high mountain broken plateau similar to Bolivia. Into its edges are cut deep valleys which have become the seat of sugar growing. East of it is a portion of the swampy tropics known as the Chaco, mainly devoted to cattle raising and forestry but which is being reclaimed and put into cotton. Beyond this is northeastern Argentina, including Misiones territory, Corrientes and Entre Rios. The first named is a bit of Brazil in character while the two provinces between the Paraná and Uruguay rivers are fertile but low-lying prairie land largely devoted to cattle raising and agriculture.

To the south Argentina narrows to a point on the straits of Magellan in south latitude corresponding to that of Juneau, Alaska, in the northern hemisphere, and where the climate is not dissimilar. If one thinks of Argentina as in the northern latitudes and reversed it could be pictured as a triangular country east of the Rocky Mountains, with the tip parallel with Sitka, with its widest portion extending out from the mountains as far as St. Louis, and with its other extreme as far south as the latitude of Havana. Its total north-south extent is 2,285 miles and its maximum width 930 miles. With an area of 1,079,965 square miles, it is the second South American country in area and it is the largest one inhabited by Spanish-speaking people in the world.

Argentina is a white man's country, almost the whole of the nearly 12 million inhabitants belonging to that race. About 20 per cent are of Italian descent, but the remainder are almost entirely of Spanish stock. The Indians who once roamed over its plains were similar in character to our own aborigines and the same long-drawn-out war of extermination took place in Argentina as in the United States. Argentina is the striking exception among South American countries in having no considerable admixture of native blood in its racial make-up. This is probably one reason why there has been important immigration from Europe through the past three-quarters of a century. Argentina as it exists today is economically a new country. First effectively settled in 1560 to 1580, Argentina had by 1850 a population of only about a million, but in the next eighty years it increased to over 11,000,000, giving a population density of about ten to the square mile. The people, however, are very unevenly distributed. Two-thirds of them are in the 250,000 square miles of pampas, where the city of Buenos Aires alone contains as many people as live in the more than 800,000 square miles of area outside the pampas. Its growth was retarded while it remained (until 1810) one of the Spanish colonies in the New World, because as an agricultural country it had no real place in the Spanish Colonial system. Administered for many years from the West Coast, all its trade with Spain, incredible as it sounds, was forced by royal decree to pass up over the Bolivian plateau, thence down to the Pacific coast and by ship to Panama, where it was transferred across the isthmus to ship again. As the natives of the region were warlike, they not only were not available as slave labor on plantations, but also harried the freight caravans, and it was not remarkable that the development was slow despite the remarkable suitability of the pampas region for stock-growing. Immigration into the country was prohibited by Spain, which had no interest in promoting foodstuff production in the New World, so the population increased only slowly. After 1810, these handicaps were removed, but there followed fifty years of internal political strife, and it was not until President Sarmiento began the modernization of Argentina in 1868 that progress became rapid.

There were a number of factors in starting this period

of rapid growth. The war-like natives were finally subdued, railroads were built radiating from Buenos Aires, Scotch, English and Irish sheep growers were allowed to come in, and were in turn followed by an influx of immigration from Southern Europe. Our own Civil War presented Argentina with European markets for her agricultural products and the introduction of refrigeration for meat made its export possible. All these, and other factors that we need not stop to mention, set Argentina upon a highway of national development that quickly led it from a minor place among South American countries to the leading place as a foodstuff producer, and an important one among the countries of the world that have a large exportable surplus of meat, wheat, and wool. This is now supplemented by yerba mate from the Misiones territory and quebracho extract from the northern forested region known as the Chaco, while the domestic economy is enriched by the sugar, wine and goats of the western highlands, and the forest products of Patagonia. Argentina has a sufficient variation in climate, elevation, and character of soil to fit it for the production of a great variety of commodities of an organic nature. What this has meant in national economy is indicated by the following table:

Exports of Cereals, Argentina, Tons

	1878	1888	1898	1908	1918	1929
Wheat	2,547	178,929	645,161	3,636,294	2,996,408	7,348,000
Corn	17,064	162,037	717,105	1,711,804	664,685	5,637,000
Linseed	104	40,223	158,904	1,005,650	391,382	1,800,000
Oats	..	..	1,107	440,041	542,097	478,000

Exports of frozen and chilled beef, valued at $12,400 in 1894, had increased to $66,640,434 in 1914 and $86,700,000 in 1929; exports of canned meats increased from $244,366 worth in 1894 to $3,893,867 worth in 1918, and $16,000,000 worth in 1929. With a total foreign trade of about $81,000,000 in 1878, and a negative trade balance of over $6,000,000, it advanced to a total foreign trade in 1919 of $1,715,200,000 and a favorable balance of $326,000,000. The foreign trade in 1928 was slightly larger than in 1919; in 1930, it declined to $1,129,000 with an unfavorable trade balance of $104,-500,000. In 1932, imports were valued at $215,034,000 and exports at $331,135,000.

Evidences of the internal changes reflecting this development are seen in the change from the importation of about 2,500,000 pounds of flour in 1880 to a domestic production of 2,000,000 pounds in 1913; an increase in sugar production from 92,000 tons in 1893 to over 200,000 in 1913; an increase in the value of wool production from $32,000,000 in 1895 to $132,000,000 in 1917; and an increase in the exports of butter from $5,330,000 worth in 1903 to $18,969,000 worth in 1918.

A better index for our purposes is the growth of the Argentine railway system, which, beginning with six and one-half miles of line in 1857 and total receipts of $19,-185, had increased to over 21,000 miles of line in 1931, a considerable portion owned by English companies and most of the remainder being government-owned. There were 360,000 registered automobiles in 1930, but only 131,697 miles of highway, of which only 500 miles were hard-surfaced, though the 850-mile road between Buenos Aires and Mendoza was expected to be completed in 1934. The relation of this to the fuel problem will be considered, but first we must turn to the general subject of mineral production.

It is evident that the period of agricultural civilization and slow growth during the three centuries prior to 1868 and the period of rapid development and modernization of the country since then differed greatly as regards needs for minerals. In the first period the mineral needs were slight except for building construction in cities like Buenos Aires. These were met, since the fine silt of the pampas region does not supply even sand for mortar, by importing of building materials from north of the river Plate and later from Europe. The small needs of metals for tools were also supplied by imports. There remained only the typical Spanish search for gold and silver. While gold occurs at a number of places it is nowhere abundant; silver is found only as a minor constituent of lead and copper ores. Copper could not be economically produced in the new world at first, and even when it became profitable the deposits found in Argentina were only of the marginal class. The same was long true of lead. Consequently, in the earlier Argentina there was neither a mineral industry, nor much need of any, though there was a good deal of search for precious metals.

Industrial development changed this situation. The building of railroads not only greatly increased the requirements of iron, but of fuel as well. The early trails from Buenos Aires to the northwest traversed or passed near the forested regions, so wood was used as fuel. In the grassy plain country fuel supply was always a problem through quantitatively the needs were small. When petroleum came to be used first for lighting and then for automobile propulsion it had to be imported, as coal had previously been. A major mineral problem in Argentina today is how to make the country less dependent on foreign sources of supply for power minerals. Progress has been made in that direction in the case of petroleum, but there seems little prospect of any appreciable success in the case of coal.

As nearly as can be estimated, the present fuel requirements of Argentina correspond to the equivalent of 6,000,000 or 7,000,000 tons of coal annually. Fuels amount to 19 per cent of the total imports of the country. Half of the national supply is furnished by importations of coal, one-sixth of it is met by the use of fuel oil partly of domestic origin and partly imported, and one-sixth by wood. In the northern forested part of Argentina wood is important for local use and charcoal is shipped out. It is also of local importance in western and southern Argentina.

Aside from wood, practically all the fuel requirements of Argentina were met, until 1914, by imports of coal, the annual amount brought in having increased from 773,870 tons in 1900 to 4,046,278 tons in 1913. During 1917 and 1918, imports fell off to 707,712 tons and 821,-974 tons, respectively,[1] because of the conditions created by the World War. Argentina typically depends on Great Britain for its coal, and at that time neither the men to mine it nor the ships to transport it were available. After the Armistice, coal imports increased, but rose only to 3,434,000 tons in 1928, declining to 3,013,000 tons in 1930 and 2,350,000 tons in 1932. To some extent, consumers in Argentina learned, as elsewhere in the world, to be more economical and efficient in their use of coal, but the decline in imports was principally due to the substitution of fuel oil for coal. One ton of fuel oil has the heat equivalent of nearly one and one-half tons of coal (3.6 barrels fuel oil = 1 ton of coal), so that its easier handling in loading and discharging ship, the less cargo space required, and the mechanical convenience of its use enabled it to compete with coal in a world where crude oil supply was more than sufficient to meet the demand for

[1] The shortage of coal was met partly by an increase in the burning of wood, asphaltic residues from oil seeps, and even of corn.

refined products. The principal users of fuel in Argentina were easily able to make the change.

Argentina is a country where the people are frugal in use of fuel. In the United States nearly 17 per cent of the total fuel consumption is used for domestic heating, in Argentina less than 4 per cent. Large consumers (public utilities, manufacturing, *etc.*) in the United States use over 40 per cent of the total fuel; in Argentina less than 25 per cent. The railroads, on the other hand, which in the United States consume only one-quarter of the total fuel, in Argentina consume nearly half of it.

Where it can be obtained at a competitive price, fuel oil is more desirable for locomotives than coal, since it can be mechanically fed to the firebox and occupies less room in the tender. Conditions in Argentina are, therefore, favorable to competition of fuel oil with coal. The importation of fuel oil and crude petroleum, which were first shown separately in statistics in 1916 and then amounted to 118,113 metric tons, had risen to 700,000 metric tons in 1923, and 1,301,393 tons in 1929; while not all of it was available as fuel oil it helped greatly to cut down the import requirements. Thus between 1916 and 1923, something like 1½ million tons of coal were displaced by fuel oil in meeting the nation's fuel requirements. Roughly speaking, the competitive line is a price of $10 per ton, delivered, for coal and $15 per ton for fuel oil; any increase in the price of coal or lowering of the price of fuel oil encourages further substitution of fuel oil for coal, and *vice versa.*

Great Britain supplies Argentina with most of its coal for a number of reasons. One is that British coal is high-grade, consequently the freight cost on a unit of heat is less. British coal producers push their coal harder on foreign markets than do others. Much of the industry (especially the railroads) in Argentina has been financed with British money and is under the technical supervision of the British. Thus the industries have been provided with equipment designed for burning British coal, and the natural tendency of those in charge of the equipment is to use British coal. Coal purchases are made through a London office, where the force of pressure to "buy British" could not be easily overcome. Finally, the large amount of shipping plying between Argentina and Great Britain makes for competitive conditions in carrying rates and ensures a steady supply throughout the year round without maintaining stocks in Argentina to provide for irregularities in arrivals.

The price of coal delivered in Buenos Aires, which ranged around thirty Argentine gold pesos per metric ton in 1920, had declined to ten to eleven pesos in 1924. Early in 1931, the c.i.f. price of British steam coal, in cargo lots, was $7.30 per ton. The price of coal is therefore not excessive, and though it might seem that the approximately $20,000,000 which the annual coal supply of Argentina costs would be a corresponding gain to the country if it could be produced at home, that assumes that labor now unemployed could be used in mining and transporting it, and there would be no interest and amortization charge on the facilities that must be provided for its production. The real question is whether it would cost Argentina more to produce its fuel than it would to buy it.

Argentina has long had a "Dirección General de Minas, Geologica, e Hidrologia" in its Ministry of Agriculture, with a competent staff, and enough information is available to make it clear that the regions that contain the most people and are best supplied with railroads do not contain at any accessible depth, strata that are at all likely to contain usable coal. The claim that coal has been found in all the sedimentary strata of Argentina from Carboniferous to Tertiary, inclusive, mostly rests on occurrences that are so far only of museum value. All the evidence indicates that coal of much value, if it is found at all, will be in beds of Gondwana, or of the Tertiary age. The deposits in these strata that have been studied have mostly proved disappointing.

Typical of the attempts to produce coal is that in the province of Mendoza, where in the period of high prices, 1917–1920, some 3,000 tons or more was taken out and about 1,200 tons shipped to Buenos Aires. The coal beds there, which are in the Gondwana, were found to be lens-shaped, irregular in thickness and extent, and so crushed, broken and twisted as not only to make mining difficult and expensive but to render it impossible that much lump coal could be produced. In addition, the coal tends to slack into fine particles when exposed. This with the only moderate heating value of the coal, its high ash content, and its distance from the main centers of consumption, makes it seem unlikely that this deposit will ever be of more than local importance. In the province of San Juan, adjoining Mendoza on the northwest, Gondwana outcrops. Near the railroad station Marayes coal has been opened in a basin of gentle dip. The beds are not much faulted or crushed and the coal as exposed is regular in thickness and well situated for mining. It, however, contains too much ash and slate to be usable. Tests indicate that it could be washed to yield a product of good quality though at a corresponding increase in cost. The field also suffers from being at a long distance from any considerable consuming center and probably is of most value to the nation as an emergency reserve.

Scattered outcrops of Tertiary coal-bearing strata are

found along the eastern front of the Andes in the south half of Argentina, but the known areas of rocks at all likely to contain coal are not extensive. Perhaps the most promising one, commercially, is near Lake Epuyén, in the Territory of Chubut. There is a possibility of the presence here of as much as 10,000,000 tons occurring in workable beds but the detailed geology remains to be worked out. The coal is sub-bituminous and low-grade. Tests made in the laboratories of the Dirección de Minas indicate that washing and preparation for the market would require about two tons of Epuyén washed coal to give the results in practice achieved with imported coal. Four to five tons of the coal would have to be mined and washed, and two tons of washed product transported, so effective competition with a ton of imported coal except in the immediate region of the deposit seems remote. The other reported occurrences in Chubut and Santa Cruz to the south labor under the further handicap that they are in a region that as yet has not much need for fuel, where wood is available, and where water power can be developed. On the whole, the occurrences cited seem representative for the whole country.

Since water is a mineral it deserves to be included here of its own right, as well as its indirect bearing on the fuel problem. The great present market for any power is Buenos Aires and the coast cities which are not only a thousand miles distant from the waterfalls on the east slope of the Andes in Patagonia but are separated from them by an arid region that offers no promise of local development. The most promising water power is that of the Iguassú Falls on the northeastern boundary, where there is an effective head of 230 feet and it is estimated that 325,000 horse-power can be developed. These falls, unfortunately, are 800 miles from Buenos Aires, and while they have been the subject of considerable engineering study the long-distance transmission involved has so far prevented anything beyond initial consideration of the project and an agreement among Brazil, Paraguay, and Argentina as to relative rights. The potential water power of the Pilcomayo and other rivers of the far northwest is even farther from commercial exploitation, the net result being that the present developed water power in Argentina is reported as only 35,000 horse-power, or about one-tenth that of Massachusetts. No one can say that water power will not eventually be of considerable importance in Argentina, but the present outlook is not bright.

As regards petroleum, both the present status and the outlook for the future are much more favorable. As early as 1830, oil seeps were recognized in the foothills of northern and central Argentina. It was not, though, as a source of power that petroleum was first developed, for the Compañia Mendocina de Petroleo began operations in November, 1886, when the only function of petroleum was still to furnish light. This pioneer attempt to develop petroleum in Argentina was not destined to be permanently successful, for although the company drilled wells at Cachetutu and laid a thirty-mile pipe-line to Mendoza, where the crude was refined, the original company failed in 1897. Various subsequent reorganizations later attempted to carry on, but without success, chiefly because the amount of petroleum obtained per well was too small. The occurrence is of scientific interest because the petroleum is found in Rhaetic beds, and it has been suggested that it was produced by the effect of hot intrusions on a black slate, but it seems of little commercial importance and is chiefly of interest in showing that as long ago as 1886 there was interest in developing a domestic petroleum supply. The form of occurrence is one that in Mexico proved to be of great economic significance and volcanic plugs cutting up through shale beds and showing oil seeps around them are found at more than one point in western Argentina.

With the development of automobiles and internal combustion engines in general, a fresh incentive for the development of a domestic supply was furnished. Fortunately, a water well being drilled by the Dirección de Minas at Comodoro Rivadavia (Chubut) about 850 miles southwest of Buenos Aires, showed the presence of heavy petroleum, of 17° to 20° Beaume. Additional drilling disclosed oil-bearing sands at depths between 1,200 and 2,000 feet and the major oil fields in Argentina came into being. Drilling is with the rotary and production costs are low. The beds, of upper Cretaceous age[2] are reported by the Argentine geologists to be almost flat, to constitute a broad shallow syncline, and are claimed to extend for great distances. The occurrence of oil in them is irregular, its distribution being perhaps controlled by differences in the porosity of the sand.

Shortly after the discovery of petroleum at Comodoro Rivadavia in 1907, the Yacimientos Petroliferos Fiscales was set up to develop the field which had been promptly reserved from private entry. Its form has changed slightly from time to time but essentially the organization consists of a President and six directors nominated by the President of the Republic and approved by the Senate, and reporting to the Minister of Agriculture. The Y.P.F. has full charge of all the petroleum reserves of the country, of which there are now several, and from three of which, the original Comodoro Rivadavia, the

[2] F. G. Clapp: "Petroleum Resources of South America," *Trans. A. I. M. E.*, vol. 57, pp. 914–963.

Plaza Huincul, and the Northern there is production. The organization operates as nearly as may be like a private company controlling everything from the initial geological and geophysical surveys in the field to the delivery of gasoline to the motorist waiting at the curb pump. In December, 1932, it celebrated the twenty-fifth anniversary of the discovery, by means of a petroleum exposition at Buenos Aires where were exhibited the original drilling rig with which the first oil was found, alongside of the super-gas plant which was then the latest addition to the facilities of the organization. At that time the Y.P.F. had 1,319 producing wells, four refineries, a fleet of tankers, a network of distributing stations covering much of the Republic, and was supplying about 25 per cent of the fuel oil, 26 per cent of the gasoline and 14 per cent of the kerosene employed in the country. According to Dr. Antonio de Tomasso, then Minister of Agriculture, the Y.P.F. had that year shown a profit of 3,000,000 pesos and had in addition reduced the cost of petroleum products to the army, navy, and national railways by the amount of 2,500,000 pesos more. The original appropriation made to the organization in 1907 had been 500,000 pesos. To this various sums had been added, until at the end of 1912 the total stood at 8,655,240 pesos. From then on the funds grew by plowing in profits until at the end of 1931 the total capital was recorded as being 259,931,472 pesos.

Argentinians are proud of the Y.P.F. and point out that it has changed Comodoro Rivadavia from a spot on the desert having neither water nor inhabitants to a modern industrial community with excellent buildings, schools, hospitals and all those facilities which large-scale industry finds it in its own interest to supply when development is undertaken in such situations. In addition, the nation undoubtedly has a feeling of greater security from holding in its own hands important oil fields in a country where other fuel is all but non-existent. The fact that this has been done on small initial appropriations, and a great modern and well-equipped enterprise created in a country where experience with industries other than those related to agriculture is almost altogether lacking, is important. It may be pointed out, however, in all fairness, that the speed of development has been notably slower than has frequently been accomplished by experienced companies able to bring to bear upon the work all the resources of knowledge and skill of the past. It is also fair to say that the Y.P.F., as is true generally of government operations everywhere, has an important advantage over competitors in exemption from taxes and in other particulars. Finally, the claim made for the Y.P.F. that it has a higher ratio of producing wells to total drilled than the average in the United States, the largest producer, is countered by the criticism of experienced oil men to the effect that it merely has not kept up development. The larger number of exploratory wells which, in their judgment, should have been drilled, hang over its head in the form of a deferred charge still to be met.

Interesting as is this experience with government ownership and operation, and with most generous recognition of the many achievements of its staff, it is by no means clear that Argentina would not have been even farther along on the road to fuel independence had judicious bargains been made with companies already experienced in the business of petroleum production. It should be further noted that the controversies that have arisen between the Republic and certain foreign companies have not been over the competition of the Y.P.F., but over a nationalization program which at times took the form of canceling what the companies considered to be vested rights. This, indeed, has in most cases been the critical matter in controversies between South American governments and foreign oil companies. It has already been indicated that there is a fundamental difference in point of view as to what is a vested right, which enters into such controversies.

In the extreme northwest there is a southern extension of the Bolivian petroleum districts into the provinces of Salta and Jujuy. Redwood[3] has described seven districts where petroleum occurs in this region, but only two, San Pedro and Lomitas, near the Bolivian border, were producing in 1931, the total output being around 2,500 barrels. These two fields were discovered in 1924, but only became producers in 1928. The oil is of high quality, 40° to 45° B. and gives as high as 70 per cent white products in refining. The crude is shipped by combined pipe-line and rail to the seacoast, where it is refined in the Standard Oil refinery at Buenos Aires. Drilling is difficult, as the strata are much folded, but the depth of the wells is only 1,100 to 2,500 feet.

Farther south, in the provinces of Neuquen and Mendoza, Arnold[4] has described eight localities at which petroleum apparently occurs, but only in Neuquen has it become important. Beginning forty miles north of Mendoza and extending south some 300 miles to latitude 42° in Neuquen, petroleum is variously reported to occur in lower Jurassic, upper Jurassic, Cretaceous and Eocene strata. Arnold thought that not over 300 square miles indicated real possibilities. About 1927, some good wells were drilled in Neuquen near Plaza Huincul, and in 1931

[3] Boverton Redwood: "Treatise on Petroleum," 3rd E., pp. 100–106, 192–196, 1931.

[4] Ralph Arnold: "Conservation of Oil and Gas Resources of the Americas," *Econ. Geology,* vol. 11, pp. 203–222, 299–326, 1916.

production there reached 5,300 barrels daily from Jurassic strata at depths of 1,500 to 3,000 feet. At this point the oil-bearing structure is, though narrow, fifteen to twenty miles long. The crude oil is shipped by rail to the refinery at Bahía Blanca, and the products thence by steamer to Buenos Aires. Near Tupungato, in the Province of Mendoza, a new field was discovered in 1934 which may prove significant.

There are seven petroleum-refining plants in Argentina. Of these, three, with a total capacity of 15,000 barrels daily, are operated by the government agency, Yacimientos de Petroleos Fiscales, and one, with a capacity of 5,000–6,000 barrels per day, by the Royal Dutch-Shell Company. The Standard Oil Company of New Jersey has three: a 10,000-barrels per day plant at Campaña; a 3,000-barrels per day plant at Bahía Blanca, and a small plant in Northern Argentina. As the total capacity of these plants is 50,000 barrels daily, and the total production of crude in 1931 was only 31,000 barrels daily, the Standard operates its main refinery on crude brought from Talara, Peru, in tank steamers. The combined output of all the plants is about half enough to meet the present petroleum consumption of Argentina. A high tariff on imported petroleum products offers a strong incentive to develop domestic supplies, but private companies can no longer obtain concessions of territory as they did before 1924. The production by private companies, mentioned above, is all from concessions obtained before that time.

It should be pointed out here that two aspects of the national policy toward petroleum run counter to each other. The tariff on petroleum products of outside origin would tend to encourage foreign companies to spend money exploring for petroleum in Argentina, in the hope that even with somewhat higher costs for production and transportation than apply to their present sources of supply, there would still be a margin between their costs delivered to consumers as compared to similar costs on imported petroleum. But since private companies no longer can obtain concessions, this force is not operative. This leaves only the Yacimientos Petrolíferos Fiscal with incentive to continue exploratory drilling. Basic conditions are unfavorable to this, water and fuel being scarce in some regions, others remote from transportation facilities, and in others finding oil is difficult because of the disturbed condition of the strata. Even in the Comodoro Rivadavia district, where none of these handicaps applies, the government has lagged behind the private companies in keeping up development.

The tariff, in itself, furnishes no incentive to a government petroleum company beyond raising the price level at which imported petroleum can be sold. Once this is fixed, the only way in which a government agency could benefit would be by increasing its operations so as to get a larger share of the total petroleum business, and a corresponding share of the profit that may be in it. This assumes that the government agency is able to make a profit on its operations at the price level set by the tariff.

So far as mineral sources of power supply are concerned it is clear that Argentina has almost no possibility of becoming self-sufficient as regards coal requirements. As to hydro-electric power, that which is available is situated at such great distances from the principal markets for power that it is not at present available. While it is possible that uses for hydro-electric power may be developed near the places where it is available, this would have no effect on existing requirements, but would stimulate additional domestic industry. As regards petroleum, a domestic supply equal to about half present domestic consumption has been developed. It seems probable that the domestic supply can be increased, but whether it will be increased faster than the rate of growth of consumption cannot be foreseen. It seems improbable that Argentina will ever have a large exportable surplus of petroleum, though by developing its local supplies to the full amount necessary to meet domestic needs it could all but solve its foreign exchange problems.

Iron, as has already been stated, is the great master metal in modern industry. In considering the situation in Argentina it is desirable to keep in mind that its production requires two steps: (1) the making of metal from ore, and (2) the manufacture of usable articles from steel, cast iron, and wrought iron. This second step requires that the first shall have been taken, but not necessarily in the same place or even the same country. In modern industry objects of steel, cast, and wrought iron break, wear out, or become obsolete, and are then sold as "scrap" to metallurgical plants, which rework them in usable form. Wrought-iron scrap is frequently made into crucible steel, cast iron may be either recast or made into steel, and steel scrap may be made into steel castings or remelted and rerolled into a grade of steel for which there may be a market. In making steel some new "pig" iron is ordinarily used, even where scrap is abundant, since it is the easiest and cheapest way to do it. Steelmaking operations are fairly flexible, and steel may be made either wholly from new pig iron, from scrap, or from a mixture of the two, the choice depending on the price of scrap, of pig iron, and the cost of steel-making operations with different proportions of them.

It can be said at once that Argentina has not the ore to support blast furnaces. From the early days of the country its requirements, then chiefly for cutting tools and agricultural implements of iron and steel, were met by the importation of the finished articles. These, when they broke, wore out, or became obsolete, were thrown aside and gradually accumulated into scrap piles. The amount of scrap was of little importance until railroads developed. Then large tonnages of steel equipment, rails, *etc.,* were imported and in time found their way to the scrap pile. This in time led to the development of a small but important industry in the remelting, recasting and reworking of iron and steel. Argentina also became an important source of scrap for European steel makers.

This situation is not entirely satisfactory from either a commercial or military standpoint. The constant purchase of new steel products made in part from scrap shipped out of the country involved double payment of freight charges, while the absence of any adequate local source of steel, other than stocks of finished goods, might well prove a source of weakness in case of a prolonged war. For a short period a country can wage war on accumulated stocks. For any long contest, as well as for maintaining a satisfactory, balanced national economy, it is necessary either to develop local supplies or to keep control of the sea so as to assure an unbroken line of imports. The authorities of Argentina have given much thought to these questions and the Dirección de Minas has made active search for the necessary raw materials.

Production of new iron is based upon: (1) an adequate supply of iron ore of good enough grade to permit working costs that compare favorably with those of competing countries; (2) the proper fuel supply, which normally means coal of a quality that will yield good coke, and (3) proper flux. The third is a requirement that is usually met without much difficulty, but the first and second are more difficult. An iron industry may be based on imported coal, which is coked locally and the gas and by-products sold, but this complicates the operation, and only partly meets the strategic problem of providing domestic resources. Even less satisfactory is the use of electric power for iron reduction, since only half of the fuel requirement is met in that way and the carbon necessary for reduction of oxide to metal must still be provided either in the form of coke or charcoal. It is evident that there is no present hope of an iron-making industry based on domestic coke and the only alternatives are the use of coke made from imported coal or the utilization of electric power and charcoal in the northern regions where both are available. The technology of making steel from sponge iron, which permits the use of inferior coal for fuel, has not yet been worked out to the point where it is competitive with blast furnace work.

Development along any line would require an adequate supply of good iron ore, but all the effort devoted to the search for the latter has so far failed to discover ore bodies of significant size and value. Iron ore has been found, but nowhere of both the quality and amount necessary for an economic development. A bibliography on iron ores, prepared by Robert Beder of the Dirección de Minas in 1924, lists forty-five reports on iron-ore occurrences, and is accompanied by a map that shows over eighty places at which it had been reported. It would require too much space to consider these in detail, and what follows is merely a general review.

Magnetite is found at various places, especially in northwestern Argentina. In the Romay district of Catamarca the Sud Americana Company at one time took preliminary steps to initiate a smelting enterprise, but did not proceed with it. The magnetite, as is usual, has a large content of titanium oxide, which would lead to smelting difficulties, and is usually irregular. The deposits are also usually distant from centers of industry and many of them difficult of access. They offer no encouragement for starting a smelting enterprise.

Magnetite sands are found along the seacoast at Mar del Plata and elsewhere and have attracted attention. Since magnetite is heavier than the sand, the sorting action of the waves tends to concentrate it, as does the wind in the formation of sand dunes. These sands have been studied with enough care to indicate that they exhibit the characteristics of similar deposits at many other places elsewhere in the world where considerable sums have been spent in fruitless attempts to develop iron-smelting enterprises on them. The striking contrast between the black magnetite and white sand invariably causes the casual observer to overestimate greatly the amount actually present, while few people can visualize the amount of ore required for a modern blast furnace, over 2,000 tons daily. The Argentine sands, as is characteristic of similar sand elsewhere, have proved to be high in titanium, another handicap to their use.

Hematite is the great source of iron supply of the world, since in many places it occurs in almost pure deposits containing many millions of tons. Unfortunately, this is not the case in Argentina. The many places where hematite has been reported, chiefly along the eastern slope of the Andes, have so far proved to be too small to be commercially useful, even enough ore to supply a blast furnace for a single day being a notable occurrence. In addition, the ore is often too high in silica to be commercial, while the deposits suffer under the handicaps of remote situation and other unfavorable working conditions. The general conclusion from all the studies made is that there are no known workable deposits of hematite in Argentina.

Residual ores, or "laterites," are present in the Misiones territory. They represent the iron left behind in the decay and removal of surface rocks. Ores of this type were mined for a while in northern Cuba and are known in the Philippines, Celebes, and other tropical localities. They are not now used for making iron, but are regarded as of considerable possible future importance. They are typically high in moisture content, and often contain constituents that create grave technical difficulties in their utilization. The deposits known in Argentina are small and shallow. Their silica content is high and they are not a promising source of ore.

The general conclusion is that the iron-ore deposits of Argentina are too small, too low-grade, and too unfavorably situated to offer hope of economic development, even if an abundant supply of suitable fuel and power were available. In the absence of the latter their status seems nearly hopeless, even for a small uneconomic enterprise that might be supported with a grant of government funds because of its strategic importance. Fortunately better means of solving the problem are available.

The opportunity for steel manufacture in Argentina lies in the remelting and reshaping of material which has reached its limit of usefulness in the form in which it had been imported. Such remelting is a large and growing business in many countries, such as Italy and Japan, each of which is poorly endowed with raw materials for making steel. Despite the immense loss resulting from rust of steel and iron, there is a steadily mounting stock. The portion of this which is recoverable, which in turn depends largely on its form and distribution, is easily available and is a preferred source of supply for new steel. Either alone or with an admixture of pig iron or hot metal in a percentage determined by relative price, it is the common basis for making open-hearth steel. In the Pacific Coast states of North America a considerable steel industry was built up in advance of development of a western supply of pig iron. In Argentina the same economic process is in progress. In Buenos Aires and in other large cities there are numerous small plants for melting and recasting iron and steel scrap. The principal railway lines maintain plants for similar re-use of old material and in Buenos Aires a beginning has been made in rolling such steel into rods, bars and small shapes. With a tariff on imports and the reëstablishment of the embargo on export of scrap, it would be possible for the country to build up a growing steel industry. It would still be necessary to import iron and steel since the amount of recoverable scrap is always less than the total originally put into use, and since the needs of the country may be reasonably expected to continue to grow. This, however, could be brought in in the cheapest form of pig iron in part and supplemented by imports of finished goods for which the local market does not warrant erection of plants. It is to be remembered that steel is used in many forms only a part of which could be advantageously supplied by local plants, most of which would necessarily be small as compared with those abroad. The advantage Argentina derives by purchasing from countries where mass production results in low costs should not be overlooked. True economy would undoubtedly lie in continued purchase of large amounts of iron and steel, but to the extent that material already in the country can be reworked the problem of foreign exchange would be simplified. The process of starting with steel making rather than with iron production has the further advantage that there is opportunity for a wide choice of fuels and even, if that be available, for use of hydro-electric power. The steel maker is not restricted, as is the operator of a blast furnace, by the necessity of having coking coal available.

Many of the non-ferrous and non-metallic minerals are found in Argentina, but no mines of major importance are now worked. There is a long record of exploration and attempted development, and the government has been unusually liberal in its support of such efforts, but so far without adequate return. The most promising field for such search is in northwestern Argentina where from La Rioja to Jujuy there are many of the geological features which characterize the mineral districts of Bolivia and Peru. Farther south, in Mendoza especially, there are deposits of unproved value at a number of points and in the Sierras of the plains, San Luis and Córdoba, with a different type of geology, various small ore bodies have been developed. In the first district Jujuy supplies lead and silver from ores similar to those just to the north in Bolivia.

Lead production in Argentina, though not large, is of considerable commercial importance, for the domestic consumption of lead, as metal, ranges in times of normal prosperity from 10,000 to 14,000 tons, as nearly as can be estimated. Of this annual requirement about 2,000 to 3,000 tons are available as recovered secondary metal, and about 6,000 tons are imported as metal, leaving 6,000 tons to be supplied from domestic production. The reported domestic production of 7,700 tons in 1925, 9,000 tons in 1930, and 8,148 tons in 1932 probably includes secondary metal.

The most important domestic producer is the Cia Minera y Metalurgica Sud America, in which the National Lead Co., S. A., (a subsidiary of the National Lead Co., N. Y.), acquired a controlling interest in 1922. It operates a lead smelter and refinery at Buenos Aires and produces ore from a group of mines which it owns in Jujuy province. Its output of refined lead in 1928 was reported as 5,205 tons. At Tres Cruces, in the same province, is a lead deposit owned by the Compañia Minera Aguilar, S. A. (a subsidiary of the St. Joseph Lead Co., N. Y.) which has been explored and for which plans were made to erect equipment for treating eight hundred tons daily of ore averaging 12 per cent lead, 14 per cent zinc, and with ten ounces silver content. It was planned to smelt the lead concentrate in a plant to be erected in Buenos Aires and to ship the zinc concentrate to Europe, but the financial outlook in 1929 caused the execution of these plans to be postponed. Imports of lead ore into Argentina were reported as 6,600 metric tons in 1928, 7,892 in 1929 and 6,500 in 1930. Practically all this comes across the border from southeastern Bolivia from small mines in the adjacent parts of that country. With sufficient tariff protection, Argentina's lead requirements could be met from the smelting of ore derived from its own territory or from Bolivia. If the amount required yearly and the world price of lead did not vary it would be easy to fix import tariffs at a level which would produce this result without making the domestic price of lead unduly high, but the variations in price and in annual requirements that actually occur make it difficult to attain in practice.

Copper production in Argentina has a long history but has not yet attained any great importance and has been decreasing, as shown by the following statistics taken from Economic Paper No. 1, United States Bureau of Mines, and probably based on estimates made by Merton & Co., London. Total production between 1851 and 1900, inclusive, 18,981 short tons. Total production between 1901 and 1905, inclusive, 1,642 short tons.

Year	Short Tons	Year	Short Tons
1906	118	1912	331
1907	246	1913	110
1908	220	1914	331
1909	661	1915–1925	0
1910	331	1926	331
1911	1,102	1927	0

In 1929, an output of 148 tons of copper ore was reported.

One of the most ambitious attempts to mine copper ores was at Capillitas in the province of Catamarca, near Andalgala. There are three mines, the Restauradora, the Carmelita, and the Grande; the first of these was opened in 1860. The typical ores are pyrite, tetrahedrite, chalcopyrite, galena, rhodochrosite, and a little sphalerite; in other words, they contain copper, lead, silver, gold, iron, manganese, zinc, antimony, *etc.,* of which only the first four are of recoverable value. In 1901, these properties were taken over by the Carranza-Lafone Copper Smelting Corporation of London, organized that year with a capital of £600,000 and reorganized in 1909 as the Capillitas Consolidated Mines, Ltd., which in 1915 wrote the capital down to £230,000. The company constructed an aerial tramway twenty-two miles long to carry the ore to small smelting plants on the plains at the foot of the mountains. These plants, which were a dozen miles by cart road from Andalgala on the railway line, have long since been dismantled. The mine is in an area of metamorphic rocks penetrated by an andesitic boss and the geology is not unfavorable to the occurrence of important ore bodies. There are extensive old workings and in 1925 there remained to be seen in them individual faces of apparently excellent ore. Despite the long record of business failure here the region is one which may still produce a mine. No figures of metal production of years are available, but the value of the total production of copper, lead, silver and gold from the beginning has been estimated to have approximated $3,000,000.

In the province of Rioja, near Famatina, another determined effort has been made to develop a mining district. There are here many copper-bearing veins containing gold and silver in a thick series of Paleozoic slates. Lead deposits are also reported to be present. The most important workings are at elevations of 15,000 to 16,000 feet and about twenty-two miles from the end of the branch railroad of Chilecitos. In 1903, the Famatina Development Company, with a capital of £400,000 took over some of the mines from the Famatina Copper & Gold Syndicate, Ltd., and undertook to develop them on a large scale. In 1907, the Argentine government built a twenty-one and one-half-mile aerial tramway from the railroad to the mines, the difference in elevation being over 11,000 feet. A smelting plant was built, but operations proved unprofitable, and the enterprise, after being reorganized in 1912, with £800,000 capital, as the Famatina Co., Ltd., is now owned by the Corporacion Minera de Famatina. It is not operating, having been shut down again after the latest resumption of operations, in 1927, under government supervision with the aid of credits granted by the Banco de la Nacion. While the ore is similar to that in the great Cerro de Pasco mine in Peru, it contains much less silver, the veins are narrow, the ore shoots short and the ore is not sufficiently high in grade to overcome this handicap. These factors, together with those of high altitude, scarcity of wood and water, and difficulties of housing workmen, made the property unprofitable. In 1911, the company produced 893 tons of copper.

At San Antonio de Los Sobres, in Los Andes province, there are other narrow veins carrying copper, lead, silver, and zinc which were worked from 1899 to 1903 by the Concordia Co., Ltd., and then taken over by the Compañia Minera la Concordia which, about 1906, constructed a hydro-electric plant, concentrating plant, *etc.*, but did not meet with financial success.

In Córdoba province, at Calamuchita, sixty miles southwest of Córdoba, the capital of the province, there are deposits of oxidized copper ores which the Rosario Co., about 1900, attempted to mine and smelt, without financial success. In 1900, the Mining Exploration Co., Ltd., attempted to operate the La Choicas and La Victoria mines in Argentina from headquarters at Tinguiririca, Chile, but without success. In Salta province there are veins that contain copper, lead, and zinc.

Broadly speaking, some copper can be found in almost every province along the eastern Andes north of Mendoza, and during the early part of this century numerous attempts to work deposits were made in various places. None of the attempts resulted in financial success and, so far as now known, copper deposits in Argentina belong to what may be characterized as the old-fashioned type, in which fairly high-grade ore is recovered at considerable expense from narrow veins. The increase in world copper production during the past quarter century has come mainly from the large-scale development of large low-grade deposits that can be worked cheaply. No deposits of the latter type are known to exist in Argentina, and until the ore reserves of deposits of this type are exhausted it seems improbable that copper will be produced in Argentina in any important amounts.

None of the known silver deposits in Argentina is large enough or rich enough to be worked for silver alone, and such silver as has been produced has mostly been a by-product of lead or copper smelting. In 1887, the Soc. Francesca de Minas of Fundicion de Nonogasta built a small smelting plant at Nonogasta, Rioja, to treat dry silver ores, but later altered it to a copper smelting plant. The following statistics of silver production in Argentina are taken from Economic Paper No. 8, United States Bureau of Mines.

Years, Inclusive	Production, Troy Ounces	Years, Inclusive	Production, Troy Ounces
1876–1880	1,625,048	1906–1910	695,087
1881–1885	1,759,226	1911–1915	375,378
1886–1890	1,342,358	1916–1920	130,300
1891–1895	3,195,397	1921–1925	118,000
1896–1900	1,516,505	1927	15,000
1901–1905	391,780	1929	15,000

Gold has been found at many places along the eastern front of the Andes all the way from the Bolivian border to Tierra del Fuego. It occurs both as gold-bearing veins and small placer deposits. It also occurs with the copper and lead ores already referred to. The following statistics of gold production in Argentina are taken from Economic Paper No. 6, United States Bureau of Mines.

Years, Inclusive	Production, Troy Ounces	Years, Inclusive	Production, Troy Ounces
1876–1880	18,970	1906–1910	30,612
1881–1885	18,970	1911–1915	19,300
1886–1890	11,833	1916–1920	6,186
1891–1895	34,538	1921–1925	16,690
1896–1900	37,333	1927	967
1901–1905	5,063	1929	1,000

Considering how long gold has been sought for in

Argentina, the small quantities obtained, and the tendency toward a declining output which the figures above show, it seems unnecessary to go into detail concerning the various districts in which it has been produced. Of recent years Argentina has contributed less than one-hundredth of 1 per cent of the world's annual gold supply. The gold holdings in Argentine fiscal institutions, estimated by the Federal Reserve Board at $607,290,000 in 1928, and at $252,698,000 in 1931, have been acquired through sales of Argentine products abroad. The chief increase was between 1893 and 1903, when there were favorable trade balances, and again between 1915 and 1928, especially the last two years of the latter period, in which they increased from $450,557,000 in 1926 to $607,290,000 in 1928. Without any domestic gold production of importance to add to fiscal holdings their maintenance in Argentina becomes purely a banking problem.

Tungsten deposits in Argentina have attracted much attention since they have been reported at a great number of places in the pampa mountain ranges, and were worked before 1900 in the Sierra de Córdoba, where the first discoveries were made. They have been described in detail in Bulletins No. 3 and 12 (Series B) of the Dirección de Minas. The only important producing deposit is the Los Condores, near Corcoran, in the Sierra de San Luis. The mineral found is wolframite, with small quantities of scheelite. The principal vein is 600 meters long and up to a meter wide. Though the value of the annual production is not great, it is of interest as a mineral of which there is at present an exportable surplus.

Tin is associated with tungsten at Mazan, in the province of La Rioja, but attempts to work the deposit have so far failed, since the ore found only contained ¼ per cent. At San Salvador, Catamarca, there is a deposit of cassiterite alone which is credited with having made a small production during 1914. The evidence so far available does not indicate that tin deposits in Argentina are likely to be of much importance. A small amount of tin is recovered from old tin cans in Buenos Aires. It would be desirable to have a domestic tin plate industry in Argentina to promote exports of preserved foodstuffs, but if it develops it will doubtless have to be based on imported tin, as is that of the United States.

A small deposit of manganese ore occurs at Piedra Parada Grande, near San Luis, from which 100 tons of ore are reported to have been shipped. Other deposits are reported to occur in the same vicinity, and in Jujuy promising deposits are known. There seems no possibility that manganese ore will be mined for export to steel-producing countries.

The arid climate in many parts of Argentina has produced saline lakes that in several places furnish a local salt supply. The South American Salt and Chemical Production Co., Ltd, has laid a four-inch pipe-line from a saline lake in the southern part of the province of Buenos Aires to its plant on the seacoast at San Blas. The waters of Lake Bebedero, near San Luis, are evaporated to furnish salt for local use, and in Chubut, two saline lakes on the peninsula of Valdez, in the dry season, yield salt around their margins. There is an abundant domestic supply available, and whether domestic or imported salt is used depends on the relative delivered price at points of consumption.

The saline lakes in Jujuy, Salta, and Catamarca frequently contain borates and are the basis of a local industry. By drying the lumps, screening them to free them of dirt, and sacking them, a shipping product containing as much as 50 per cent boric anhydride can be obtained. Probably most of it is exported; no data are available as regards domestic uses. The available supply seems to be adequate to meet all probable future needs for boron and its compounds. Extinct volcanoes in Los Andes province have deposits of alum about their craters. No data are available as to commercial production. In the region just mentioned, sodium carbonate is found associated with sodium sulphate and chloride. A small amount is produced for domestic use.

Sulphur deposits are also found in these extinct volcanoes, but attempts to work them have not been generally profitable, as they lie at high altitudes and are too remote from the regions where sulphur is consumed. They are chiefly important as a strategic domestic reserve.

Pegmatite veins containing mica have been found at various places in the Sierra de San Luis and Sierra de Córdoba. Small shipments to Europe have been made, but the available evidence does not indicate they will become important sources of exportable material. Fluorite occurs in veins near San Roque, Córdoba, and such data as are available on the deposits indicate they could be worked and furnish a domestic source of supply for a domestic steel-making industry, though probably not material which could stand the transportation cost to steel-making countries. No production has been reported.

Argentina is singular in that the more populous portions suffer from the absence of ordinary building materials, clay for making brick being the only one which is abundant. Wood, out of which the pioneers of North America built their first homes, is absent over the great pampas and little that is suitable for building is found elsewhere. Even ordinary building sand must be imported from Uruguay for use in Buenos Aires, and it required a careful search to find sufficient limestone and clay for establishing the Portland cement plant in Buenos Aires province. Near Córdoba limestone outcrops and there is a flourishing lime-making industry to supply not only the building needs of the country, but the sugar refineries of Tucumán. In the cities imported materials are employed and fine structures have been erected, but over the country as a whole houses are made of locally burned brick or adobe, with galvanized iron for roofs.

Minerals have, as a whole, played but a small part in the development of Argentina. Mineral exports supply less than half of 1 per cent of the total. Fuel and petroleum products amount to 19 per cent of the imports, machinery and vehicles, largely made of metal, constitute another 13 per cent and steel manufactures form 11 per cent. So far as can be seen, imports of all of these are likely to continue, though, as already pointed out, domestic fuel supplies can be increased by further developing the oil fields. While exact figures are not available it is estimated that about $25,000,000 originally was invested in petroleum development, with a possible $10,000,000 additional for all the other mineral industries. Perhaps 15,000 workers in all are employed in the mines, oil fields and reduction works. As compared with the output, capital invested, and number employed in agriculture, these figures are almost insignificant and they emphasize the fact that in Argentina the mineral industry is, and probably will continue to be, strictly subordinate.

CHAPTER IX

CHILE

THOUGH Chile ranks seventh in size and fifth in population[1] among the countries of South America, politically and commercially it is one of the three that lead; from the mineral standpoint it is also of major rank. One says among the first three because it is as difficult to attach absolute weight to the various factors in the mineral situation as it is for the political one. It might be said that Chile produces more coal than any other South American country and also more copper and iron ore, while it is not only the largest producer of sodium nitrate but is the only country in the world that is an important producer of that mineral commodity. On the other hand, Chile produces no petroleum and does not rank high as a producer of either gold or silver. On the whole it seems best not to attempt to be more specific.

In comparing it to Argentina and Brazil, it might be noted that Argentina has never had and, with the sole exception of petroleum, does not now have much mineral production; nor does there seem much possibility that it ever will have. Brazil, on the other hand, was in the eighteenth century, one of the two chief gold-producing countries of the world, and the leading source of diamonds. It now produces only one-fiftieth as much gold as the Transvaal, which at present leads in this field. This

[1] Area 286,932 square miles, and 1932 estimate of population 4,287,445.

change in rank has come about chiefly because gold production elsewhere has so greatly increased, though Brazil's production itself has declined to less than half of what it once was. For the same reasons its diamond production has also become relatively unimportant. But Brazil is now an important producer of manganese and has mineral deposits, notably of iron ore, that may later give it high world rank.

Chile, on the other hand, never became of much importance as a mineral producer until the middle of the nineteenth century and reached its maximum during the first quarter of the twentieth. Whether the factors that led to a sharp decline in its mineral output after 1929 will be temporary or permanent in their effects is a question that can be better discussed after its mineral industry and resources have been reviewed.

Chile is a country of peculiar shape, almost like a piece of tape stretched from about 17° south latitude to past 55°, or more than two-fifths of the distance from the equator to the South Pole. With its eastern border along the crests of the Andes nowhere more than 250 miles distant from the ocean that forms its western boundary, its profile from west to east is not simply a steep slope from sea level but is quite irregular. In the northern two-thirds of the country there is generally a range of mountains paralleling the coast and at no great distance from it. Behind this is usually a longitudinal valley separating this Coast Range from the foothills of the Andes. This valley is commercially the most important part of Chile and also contains[2] most of the important mineral-producing areas. In the southern third of Chile the foothills of the Andes extend to the ocean, and for this as well as other important reasons very few people inhabit that

[2] The Braden copper mine is in the Eastern Cordillera.

part of the country. Its mineral resources are not only little exploited but comparatively unknown. To make these reasons clear some general features of the situation must be considered.

As already briefly mentioned, the air surrounding the earth tends to slip as the globe rotates from east to west, thus producing the effect of winds blowing from the west. But the high temperature at the equator furnishes a stronger force tending to make the air rise. These two forces, tending to create air circulation in two planes at right angles, so react that in the southern third of Chile the prevailing winds are from the west and since they have passed over a wide stretch of ocean, they are moisture-laden. On meeting the mountains the winds rise to flow over them; as they rise they expand (because of lessened pressure) and the expansion cools them so that they precipitate their moisture as rain or snow.

The southern end of Chile is therefore a mountainous region with a cold rainy climate, resembling that of Southern Alaska and British Columbia. Like the latter, it is covered with forest; the trees are of great variety but various species of beech predominate. Lumber is exported, but difficulties of transport hamper development of the lumbering industry. Toward the south the heights above 2,500 feet are covered with perpetual snow. Near the Straits of Magellan the mountains die down to low-lying plains that are grass-covered and hence suitable for sheep-raising. The town of Magellanes (formerly known as Punta Arenas), with its 22,000 inhabitants, is the metropolis of this region, but aside from it and a few smaller places, Chile, for 600 miles north from the southern tip, is almost uninhabited. The average population for 1,000 miles from south to north is less than one per square mile.

The few people in this region and the vegetation-covered surface are handicaps to mineral discovery but nevertheless many mineral occurrences have been reported from this area. The only production worth mentioning is that of a few thousand tons of coal yearly near Magellanes, and some gold and platinum panned from the sand of the beaches of Chiloé island, which marks the northern end of this region. Aside from these all that can be said for the area is that our present scanty knowledge of its geology and mineral resources indicates that at some future time its mineral production may be more important than it appears at present.

The south winds warm as they pass north and take up moisture, instead of depositing it. As a result, northern Chile shares with the Sahara and part of Central Asia the doubtful honor of being one of the typical great desert regions of the world. Parts of it are absolute desert, being completely bare of plant life. Nowhere is the annual rainfall over ten inches, and in many places there is no rain for years on end. In a 500-mile stretch of coast only one small river reaches the sea, and, except for the winter season, when light west winds bring some moisture to the mountain slopes at elevations of from 8,000 to 10,000 feet, little rain falls. Beyond a few irrigated oases in the desert, the area is practically uncultivated. Thus in two-fifths of all Chile only two in 10,000 acres grow crops and one acre in forty supplies pasture. Were it not for its mineral resources this region would be practically untenanted; as it is, it furnishes the major fraction of Chile's exports since all of the nitrate and much of the copper is produced here.

Between these two extremes of a warm desert area and a cold rainy region lies central Chile. A transition zone between the region of prevailing southeast and steady west winds, it shades insensibly into the desert on the north and the rainy region on the south, extending for 900 miles between them (30° to 43° S. latitude). In addition to being the great agricultural region of the country, it is forested toward the south and is well supplied with mineral resources throughout but especially with copper toward the northern end and coal toward the southern. All of the manufacturing and most of the general commerce are in this region, and in it the great majority of the people live.

Central Chile corresponds in latitude to our own coast from Jacksonville, Florida, to Portland, Maine, but is neither so warm at one end nor as cold at the other. This is because of differences in the ocean currents. The cold Labrador current impinges on the coast of Maine and the warm Gulf Stream raises the temperature of Florida. In contrast, the prevailing west winds in the southern Pacific ocean set up an eastern current that is cooler than the Gulf Stream and warmer than the Labrador current. Reaching the impassable barrier of Chile, part of this turns south and part north. This current not only cools north-central Chile but helps make northern Chile arid, since any moisture-laden winds from the west are chilled in passing over it and tend to deposit their moisture at sea instead of on the land.

Two other general features may be noted. In spite of its high ratio of coast line to area, Chile has almost no good harbors among its more than sixty ports. The general absence of rivers flowing into the sea in northern Chile has been mentioned. Not only are the ports there poorly protected from storms and ocean waves, but their very number is in itself a disadvantage, since the commerce at any one is not large enough to warrant expensive port works. At most places vessels must anchor off-shore and transfer cargo with lighters and barges; there are docks at Valparaiso and a breakwater at Antofagasta.

Of railways, Chile has one long line, extending down the longitudinal valley from Pisagua at the north to Puerto Montt, some 2,500 miles to the south. An equal mileage is divided among many short lines that either connect the Longitudinal with coast ports or feed it from producing centers to the east; of the total of about 5,600 miles, 3,600 miles are state-owned. At Los Andes, connection is made with the railway across the Andes to Buenos Aires; this was the only place in South America where it was possible to cross from the Atlantic Ocean to the Pacific by railway until the connection between La Paz and Buenos Aires was made a few years ago.

The history of mining in Chile is of considerable interest. Reference was made in the first chapter to the fact that the Spaniards made their way into Chile from Peru. In this they were following the example of the Incas, who had invaded Chile during the fifteenth century and extended their sway as far as a couple of hundred miles south of what is now Santiago. Here the natives were too fierce to subdue, and the similar experience Valdivia had with them when he arrived there in 1541 has already been described. During the next two and a half centuries, during which Chile was under Spanish rule, no great amount of mining was done, for the region occupied was not particularly rich in the gold, silver, and quicksilver which the Spaniards desired. There was an abundance of copper, but it was so cheap in Europe that only exceptional deposits could be worked at a profit in South America. It is reported that during the 270-year period, 1541–1810, Chile produced 7,400,000 ounces of gold, 8,500,000 ounces of silver and only 83,000 tons of copper. The amount of nitrate produced is not stated, though it is believed that the Spaniards worked the deposits in a small way for the production of gunpowder.

A contemporary picture of mining in Chile in the first quarter of the nineteenth century is given in John Miers' "Travels in Chile" (2 vols., London, 1826). Miers had been told that copper of fine quality could be obtained in Chile for half its price in England, that coal cost practically nothing there, wages were one-fourth those in England, and that there was a great demand for rolled sheet copper along the Pacific coast. With this information he sent out to Chile a hundred tons of equipment for refining and rolling copper. Chapters XXII and XXIII of his book are devoted to mines and mining.

Miers describes in some detail the system persisting from early Spanish times under which a *habilitador* advanced the capital and took the risk of mining operations carried on by a *minero* under contract. The habilitador furnished the minero with steel, powder, food, clothing, *etc.*, and if the venture was a failure the former bore the loss. Miers was impressed with the skill and ingenuity displayed in working the mines in a small way with rather primitive means and concluded that it would not be profitable to invest capital in an attempt to work them after the English fashion. He estimated the copper output as 1,340 tons yearly, gold as £136,000 worth, and silver £36,000 worth, which was somewhat less than it had been before 1804. He ascribed this to the discouraging of investments in mining through the confiscation of the property of all royalists (or those who might be accused by the revolutionary authorities of being such); to crop failures in 1820–1823 which made food costs (and therefore working costs) high, and to the smuggling of gold and silver bullion, on which the government duties amounted to 27 per cent. Miers thought the prospects for the profitable investment of British capital in mining enterprises in Chile discouraging, and returned to Buenos Aires, where he took a contract to supply the machinery for a government mint. The only reference to nitrate production in Chile which he makes indicates that the production costs were so high as to be unprofitable, and he elsewhere says that the only market for lead in the country was for bullets; it was supplied by a mine near Santiago which produced about a ton of lead yearly.

It must be remembered that the 1,340 long tons of copper which Miers estimated Chile to produce annually was an important amount in the decade 1811–1820. Great Britain was then producing about 7,300 tons yearly, Sweden 690 tons, Germany 380 tons, Norway 260 tons, and all other countries 970 tons, according to N. Brown and C. C. Turnbull.[3] They estimated the world's total production of copper for that decade as 96,000 tons; it increased to 135,000 tons for 1821–1830, most of the increase being due to a gain from an average of 7,300 tons yearly in Great Britain to 10,600 yearly in 1821–1830. In 1831–1840, Great Britain's output reached 143,000 tons for the ten years, the improved equipment that had been supplied to the Cornish mines having then produced its maximum possible result.

Copper production in Chile during 1831–1840 is estimated[4] at 50,000 short tons for the ten years, or a little over 13 per cent of the world's production for the period. In the succeeding decade it rose to 98,000 tons, or nearly one-fifth of the world output. For the years after 1844, official export statistics for Chile are available. In the decade 1851–1860, Chile passed Great Britain as a copper producer, its output of 240,240 tons being 31 per

[3] "A Century of Copper," p. 12, London, 1899.

[4] Economic Paper No. 1, United States Bureau of Mines.

cent of the world total. Chile remained the world's leading producer of copper until 1881 when the United States, whose production in 1841–1850 had only been one-fortieth that of Chile, yielded 820,408 tons for the period 1881–1890, as compared to Chile's 389,681 tons. Chile, which had furnished over 43 per cent of the world's copper in 1861–1870, contributed only 16.44 per cent in 1881–1890. The relative decline was due to the great increase in production in the United States, Spain, and Germany, and to an actual decline in Chile from 513,744 tons in 1871–1880 to 389,681 tons in 1881–1890; a result of the exhaustion of rich surface deposits of oxidized ores and the greater difficulty and expense of working the sulphides.

It will be seen, therefore, that Miers' judgment was poor, and that at the time he wrote Chile was actually at the beginning of a great expansion of its copper industry. Until 1880, Chile copper was the dominating factor in the world's markets, shipments from that country and warehouse stocks being the principal influences in determining price quotations for the metal. After that time the United States supplies became the controlling influence. It may also be pointed out that the general policy of the copper smelters in Great Britain had a marked influence on the course of development. In the early part of the nineteenth century they occupied such a superior position in copper technology that they were able to dominate the situation, and the copper production credited to Great Britain was in considerable part from ores and crude metal sent there from all over the world for reduction and refining. The policy of secrecy of the British smelters prevented them from keeping in touch with the improvements being made in copper technology, while their belief in the security of their position left them unaware of the necessity for improvements in their methods. The terms on which they paid for ore and crude metal appeared unreasonable and unfair to those from whom they bought, with the result that smelting and refining plants nearer the centers of ore supply were erected. Probably this was inevitable since the transport of ore, matte, *etc.*, such great distances was uneconomic, but the development would not have taken place so quickly if the British smelters had appreciated their true situation and taken appropriate steps to meet it.

Thus, during the middle half of the nineteenth century, copper was Chile's principal mineral product, the region in which nitrate is produced not having come under Chilean jurisdiction until after 1879. The value of the copper (in ores, intermediate products, and metal) exported from Chile in the period 1843–1890 was over $456,000,000, according to a compilation of figures made by the Sociedad Nacional de Minera from the custom-house records.[5] Silver bars exported during the same period were valued at $132,252,528, and in addition some $17,000,000 worth of silver ores, and unknown quantities of silver in silver-lead ores and in copper products were shipped. Gold exports for the same forty-seven-year period amounted to only $7,953,722, of which nearly half was in gold-bearing ores that were exported between 1886 and 1890. According to "Mineral Industry" (*loc. cit.*, p. 552) the Mint in Chile coined over $66,000,000 worth of gold condors (13.7277 grams of gold each) between 1772 and 1890. The gold production in the same period was about $75,000,000 worth, most of it in the first half of the period, as the gold discoveries in Cali-

[5] "Mineral Industry," vol. I, p. 548, 1892. Exact figures are impossible as the silver value is included in some of the copper products, while some silver ores exported also contained copper.

fornia and Australia just before 1850 caused gold production in Chile to decline to one-fifth of the earlier output. The 999,880 ounces produced in 1801–1810 was 17 per cent of the world total, the 112,527 ounces for 1851–1860 was only 0.17 per cent of the world output. Production in 1929 was only 25,925 ounces, but somewhat increased in subsequent years through government stimulation.

Silver production, on the other hand, increased steadily. Estimated at only 2,500 ounces yearly for 1693–1700, it had grown to 322,000 ounces annually in the period 1811–1820. Thereafter it rose nearly as rapidly as copper production (with which it was associated) reaching its peak in 1887, when the value of the silver bars exported was $8,291,920. Since 1920, the production has been about 3,000,000 ounces yearly, or about three-fifths of the maximum.

Historically, therefore, Chile was chiefly a gold-mining country before 1800. After 1850, gold production (which had never been large) greatly declined, but copper and silver mining increased, reaching a maximum about 1880, when nitrate mining began its spectacular development. The second marked increase in copper production that began with the opening of the Braden mine in 1905 and the decline in nitrate production, which was one of the by-products of the World War, again made copper mining in 1932 the principal mineral industry of Chile.

During the first heyday period of copper production, 1830–1882, the mines were widely distributed from the northern border to the vicinity of Santiago, the region south of Santiago not then being of importance in copper production. Until 1842, high-grade oxide and carbonate ores were the ores mined. Operations were small and the ore was smelted in little charcoal furnaces. In 1842, Charles Lambert built a reverberatory furnace at Serena and in 1857 a copper blast furnace. Production increased steadily until it reached 43,000 long tons of fine copper in 1882, but declined after that, for reasons which will be discussed later.

The mining laws were liberal, mines being taxed about $4 per acre per year, and the holder of the property had the equivalent of a freehold, so long as he paid his taxes. Government records for 1903 showed 7,730 copper-bearing properties, of which 778 were actually operating; some of them were primarily gold or silver producers. Antofagasta was by then a province of Chile, and was one of the three leading copper-producing provinces, the other two being Coquimbo and Atacama. The largest mine in 1903, the Dulcinea of the Copiapo Mining Co., Ltd., originally opened in 1854, produced about 3,000,000 pounds of copper yearly from ore containing 15 per cent copper; this corresponded to 7 per cent of the total output of Chile at that time. The Central Chile Copper Co., Ltd., also a British company which operated four mines, produced 4,596,480 pounds of fine copper in 1905, was greatly exceeded in output by Sociedad Chilena de Fundiciones of Coquimbo, which produced nearly 24,000,000 pounds of copper in 1903 at its two smelters at Guaycan. Most of this, however, was from "custom" ore (bought from mines) as the company operated only three mines. The Nueva Sociedad Compania de Lota i Coronel with two smelting plants, the larger at Lota, produced 15,600,000 pounds of copper in 1903. The Société de Mines de Cuivre de Catemou, also operating two smelters, produced 4,400,000 pounds of copper in 1905. The Las Animas Mining and Smelting Co., Ltd., a British company, the Taltal Railway Co.,

Ltd., also a British company, and Santiago Vicuña, all operating in Atacama, each were producing about 3,500,000 pounds yearly in 1903, the latter two being merely smelting companies. The Compañia de Minera de Maipo, smelting ore from its own mine in Santiago, was the second largest single mine (the Dulcinea being the largest) with an output of a little less than 3,000,000 pounds in 1903. The Compañia Minera de Gatico, operating in Antofagasta, produced 2,500,000 pounds the same year, this property having just been acquired from Artola Hermanos. Besa y Co. sold their property at Chanaral, Atacama, in 1906 to the Société des Mines et Usines de Cuivre de Chanaral, a French company which increased the output from less than 1,000,000 pounds in 1903 to 2,500,000 pounds in 1906. The Sociedad Industrial de Atacama, in Atacama also, produced about 2,400,000 pounds in 1903. There were other smaller companies, but the list is long enough to indicate that there were many mines of fair size, though no very large ones, and that custom smelters handled the ores from small properties. But in February, 1933, the Société de Mines de Cuivre de Naltagua, a French company, the Compañia Minera de Tocapilla and the Compañia Minera Disputada de los Condes were the only properties, in addition to the three big mines that will next be described, that were in operation.

Some of the capital invested in copper mining and smelting was European, and the function of the copper in providing a trade balance for imports was only one of its important influences. The need for improved transportation led to road and railroad building, additional employment at increased wages was provided for the natives, as is evidenced by the complaints from each mining district when the opening of another attracted away part of its labor supply, and the shops of the plants functioned as trade schools in teaching practical arts to natives who had known no occupation other than farming. The export of copper led to the opening of ports, improved communications through the increased shipping between Chile and Europe, and brought about the development of local banks and fiscal relations with the world. In short, it was probably the most important factor in raising Chile from an obscure position to an important one in world trade. It also, in part, led up to the later development of nitrate production, which was in turn to cause further development.

The renascence of copper mining that began in 1905 was not an indigenous effect, but a reaction to conditions that had arisen in the United States. Our own country had in 1881–1885 attained the leading position in world copper production, when its output of 28 per cent of world production was more than that of all the South American countries combined. Ten years later it produced over 50 per cent of the world's supply, and steadily increased its lead to 67½ per cent of the world's total in 1906-1910. This great growth was fundamentally based on the steadily increasing demand for copper, especially in the United States, but was equally fundamentally based on the general policy of our copper producers, who not only permitted but encouraged inspection of their operations by other producers, with the result that any improvement in technique made by one producer was promptly adopted by all the others, with payment of at most but a reasonable patent royalty. Thus copper metallurgy in the United States reached a state of advancement that nowhere else in the world was even approached.

This great development had been made within a traditional framework, typically the mining of rather small veins of what would now be considered rich ore, its mechanical concentration, and subsequent smelting. But in 1899–1903, a revolution in copper production was brewing. D. C. Jackling, convinced that a large, low-grade copper deposit in Utah could be worked at a profit if the scale of operations was large enough, was then succeeding in convincing financial backers of the soundness of his views. The story of this enterprise is told in detail by A. B. Parsons[6] and need not be reviewed here beyond making the point that the success of the project created a new viewpoint in copper mining. Instead of looking only for deposits rich enough to bear rather high production costs, entrepreneurs now began to look for deposits large enough to be worked on a scale that would permit low operating costs. Several such were found in the United States, and three notable ones were discovered in Chile.

High on the slopes of the Andes, at an elevation of 7,500 to 9,000 feet, eighty miles southeast of Santiago, is a copper deposit from which Juan de Dias Correa, a large landowner of the region, had mined about 50,000 tons of high-grade ore previous to 1890, when the property was abandoned and the rights to it lapsed through failure to pay taxes. In 1897, Enrique Concha y Toro acquired it and adjoining areas, but his efforts to interest European capital proved fruitless. Enough work was done, however, to reveal that there was a large amount of ore containing 2 per cent copper. William Braden, searching for large low-grade deposits, visited the property in 1903, realized its possibilities, and was successful in raising the necessary capital in the United States. In 1904, the Braden Copper Co. was incorporated with an authorized capital of 625,000 preferred shares and

[6] "The Porphyry Coppers," New York, 1933.

an equal number of common. During 1906 and 1907, some 3,500 tons of concentrated ore were shipped to the United States for smelting. A railroad was constructed and a smelter built that produced matte containing 50 per cent of copper, which was shipped to Tacoma or Baltimore for further treatment. The production cost was about nine cents per pound. By that time it was evident that further investment of capital would lead to increased operating profits, so the Braden Copper Mines Co. was organized to take over the Braden Copper Co., of which the Guggenheim Exploration Co. had purchased $500,000 in bonds two or three years earlier. Under the new control, Braden issued $4,000,000 in 6 per cent convertible bonds, which provided about $2,500,000 in additional working capital after retiring existing bonds and preferred stock. The holding company was capitalized at $10,000,000, which was increased in 1911 to $14,000,000. A new concentrating plant of larger capacity was built, the smelting plant enlarged and remodeled, and various other expenditures incurred that made possible raising the output from 3,000,000 pounds of copper in 1908 to over 16,000,000 pounds in 1913.

In 1915, the Kennecott Copper Corporation bought 95 per cent of the shares of the Braden at a price, including bonds and accounts payable, of about $57,000,000. Thus the property, still operated by the Braden Copper Co., is not wholly owned by the Kennecott Copper Corporation. During the period 1918–1920 the property realized an operating profit of about $20,000,000 and operations were continually enlarged and improved until in 1932 the mine had a capacity to produce over 200,000,000 pounds of copper yearly—almost half as much as the whole country had produced in the decade 1851–1860, and three times its maximum annual production before the Braden mine was opened. Ore reserves on January 1, 1931, were estimated as 230,000,000 tons, averaging 2.18 per cent copper. Actual production in 1929 was 176,325,895 pounds of copper.

Meanwhile, two other large copper mines had been opened in Chile. The largest of the three is the Chuquicamata. North of Calama, a station on the railroad extending up from Antofagasta into Bolivia, and at an elevation of 9,000 to 10,000 feet in a desert region, are copper deposits that had been worked by the natives before the coming of the Spaniards. From then until after 1900 various attempts were made to work them, but without success. Lack of water and other difficult physical conditions were unfavorable to operation, the copper occurred in the form of unusual minerals that were difficult to treat, and the average tenor was low. All attempts to work the ore bodies on even a moderate scale ended in failure. The final development was also a reaction to something that occurred in the United States, where A. C. Burrage had fruitlessly spent in Montana large sums in attempting to perfect a process based on roasting sulphide ore, leaching the copper with calcium chloride solution, and then precipitating the copper.

The property at Chuquicamata had been presented for consideration to the Guggenheim interests about 1903, without result, since the methods in process of development for treating large low-grade sulphide deposits did not seem applicable at Chuquicamata. In 1910, this deposit was brought to Mr. Burrage's attention as a property to which the process in which he was interested might be applicable. His first examination revealed that it was too large for him to handle, so he reopened negotiations with the Guggenheim interests, who, in 1911, assumed control with Burrage holding a minority interest. Early in 1912, the Chile Exploration Co., with a capital of $1,000,000, was incorporated to open this mine.

By 1913, exploratory work had revealed the existence of 154,000,000 tons of ore, averaging 2.43 per cent copper, and the Chile Copper Co. with a capital of $110,000,000 of which $15,000,000 was 6 per cent convertible bonds was organized to acquire the physical properties of the Chile Exploration Co. None of the bonds was sold to the public until 1915. Immense sums were invested in developing and equipping the property, and some unexpected difficulties were encountered. Among these was the discovery that the material as mined contained nitrates as well as chlorides and sulphates, creating what at first seemed an almost insuperable difficulty in the chemistry and practical operation of the process. This was the first really large-scale mining operation involving chemical treatment (leaching and precipitation) as contrasted to the concentration and smelting processes used for other low-grade deposits. By the expenditure of large sums on extremely clever research work, all the difficulties were eventually successfully solved and in the first quarter of 1929 the property was producing copper at the rate of 400,000,000 pounds per year, at a cost of about six cents per pound delivered in Europe. The deposit is now believed to contain over a billion tons of ore averaging over 2 per cent copper, and may eventually prove to contain much more workable ore. The deeper-lying ore is sulphide, and when it is reached leaching will probably be discontinued and concentration employed.

The third great modern copper mine in Chile is that of the Andes Copper Mining Co., at Poterillos, fifty-five miles east of the Longitudinal railway line, and about ninety miles east of the port of Chanaral. Here attempts had been made to mine a copper deposit at various times since 1875. William Braden, after relinquishing the active direction of the Braden mine, became interested in it and spent sufficient money in exploration with drills so that he was able to interest the Anaconda Copper Mining Co. in it. In 1916, the Andes company was organized, with a capital of 500,000 shares, of a nominal par value of $100 each, and Mr. Braden transferred to this company all the rights he had acquired in exchange for an appropriate number of shares. More money was then spent on exploration. Wharves and warehouses were constructed at the port of Barquito, immediately south of Chanaral, and plans were made for the erection of a concentrating plant, a smelter, and a leaching plant. Altogether nearly $2,000,000 were spent on research, experiment and design work. By the end of 1924, in addition to the work just mentioned, nearly eighteen miles of drill holes for exploratory purposes had been put down, an equal length of horizontal and vertical workings for mining purposes made, nearly sixty miles of railway line built, eighty miles of temporary electric-power transmission line and a temporary power plant constructed, 124 miles of water-pipe laid to bring water from the east side of the westernmost range of the Andes, a town of 250 dwellings, hospital, schools, bakery, stores and public buildings built, and additional normal mine equipment was provided by the company all without any public offering of shares in the enterprise; altogether an investment of about $20,000,000 before beginning metallurgical plant construction. The ore reserves were estimated at 137,400,000 tons, averaging 1.51 per cent copper. The capital was then increased to permit issuing $40,-

000,000 in 7 per cent convertible bonds which were underwritten by a syndicate headed by the National City Bank.

With the fresh capital thus provided, the smelter and concentrating plant was built and mining operations began in December, 1926. In 1927, 54 million pounds of copper were produced; in 1928,[7] the output was 104 million, and in 1929 rose to 162 million. Production was decreased to 94 million pounds in 1930 and to 83,700,000 pounds in 1931. At the time the bond issue was floated, in 1924, it was estimated that this enterprise could produce copper at six and two-third cents per pound for operating costs (taxes, interest, depreciation, and depletion not included), and the published reports indicate that this figure was substantially attained. Unfortunately, the base price of copper c.i.f. Hamburg fell below six cents per pound during most of 1932 and it is now evident that the decision to start this large enterprise was based on a confidence in a higher stable price for copper than subsequently obtained.

These copper-mining enterprises of the first quarter of the twentieth century may profitably be contrasted with copper mining in Chile in the first quarter of the nineteenth century as described by Miers, and the industry at the end of the nineteenth century, as described above. The average price of English tough copper was £160 per ton for the decade 1801–1810, £130 per ton during the next decade, averaged £101 per ton during 1821–1830, £94 per ton during 1831–1840, and £88 per ton in 1841–1850. Miers, after studying the production methods used in Chile about 1820, could not see how, through the use of greater capital, lower production costs could be attained. The reduction in cost per

[7] The plant for the treatment of oxidized ores began operations in June, 1928.

ton of ore handled that the Braden, Chuquicamata and Andes enterprises attained about a hundred years later was as completely beyond his comprehension as the airplane travel that also characterized the latter period would have been.

It is worth pointing out, however, that copper mining in Chile in 1820 involved very little use of capital, the advances of the habilitador to the minero representing little more than operating costs for wages and supplies. Copper mining at the end of the nineteenth century was in part still on a small scale, but had largely come into the hands of smelting companies that had gradually built up plants out of the profits of small-scale operations. But copper mining in 1920 represented an enormous investment of capital. In the case of Andes over $35,000,000 had been spent on developing and equipping the property before the first ton of copper was produced, and $12,500,000 more was invested in additional equipment and development during the following two years. This capital was not savings from profitable enterprises in Chile, but savings made in the United States and invested in Chile in the hope that it would return a profit. Much of this capital went into the drilling of exploratory holes and the making of necessary preliminary underground excavations. The larger part of the money thus expended was paid out in wages to Chilean laborers, and a large part of the salaries paid out to foreign technical employees was also expended in Chile. Part of the capital invested in equipment represented its cost of construction and transport to Chile, but a considerable part of it represented transport and erection costs in Chile that increased materially the circulating wealth in the country. Other large sums were spent in providing living conditions for the native workmen far superior to what they had previously known; especially sanitation and all the conditions that lead to improved health.

This should be contrasted with the early mineral production, for which the entrepreneurs brought little or no capital into the country, mostly developing their enterprises out of the profits of the undertaking. About the only contribution such enterprises made to social development in Chile was the providing of additional employment. Copper production in the middle period was characterized by the investment of a moderate amount of outside capital, but its twentieth-century development was outstandingly the investment of huge sums drawn from outside Chile, and which could not have been derived from within Chile because neither the capital nor the technical knowledge that was a prerequisite to its reasonably safe investment existed within the country.

Nitrate

The history of sodium-nitrate production in Chile begins with 1880, because the region in which the deposits are found was, previous to that time, either Peruvian or Bolivian territory.[8] The relative rights of Bolivia and Chile in the region about Antofagasta had been for some time under dispute, but in 1874 an apparent settlement of the difficulty was made. The early development of the nitrate deposits was chiefly by Chileans, and when Bolivia imposed an export tax on nitrate a Chilean company refused to pay it, on the ground that it was in conflict with the treaty of 1874. The Bolivian government tried to seize the property of the recalcitrant company, and Chile sent a warship to seize Antofagasta. The final outcome was a war between Chile on one side and Bolivia

[8] L. W. Strauss: "The Chilean Nitrate Industry," *Min. and Sci. Press*, vol. 108, pp. 972–978, 1014–1019, 1049–1052, 1914.

and Peru on the other in which Chile decisively won. As a result, Chile took from Bolivia the province of Atacama, and from Peru the coast line north of the Bolivian possessions to and including the province of Tacna. Peru, by the Treaty of Ancon, in 1883 surrendered to Chile the province of Tarapacá, in which most of the nitrate deposits are, and ceded Tacna and Arica for ten years, with provision for a plebiscite at the end of that period. Long a source of friction between the two countries, the whole matter was finally settled in 1929 by awarding Tacna to Peru and Arica to Chile. With the political aspects of this matter we have here no concern; it is to make clear that until 1880 none of the then producing region, and very little of the area now known to contain nitrate deposits, was under the jurisdiction of Chile.

The economic basis of the political problem had, however, been long in development. Although plants and animals live in an atmosphere that is four-fifths nitrogen, they are unable to use it directly and are dependent on nitrogen compounds that are produced in a variety of ways but chiefly by the bacteria in the soil that find their favorite habitat on the roots of leguminous plants. By careful conservation of the circulating supply of nitrogen compounds, as in Oriental countries animal excreta is conserved and returned to agricultural land, this difficulty was met until, after the Middle Ages, a new and destructive use for nitrates was developed. This was the making of black powder, for which potassium nitrate is required. The typical way of obtaining this was then by piling up manure and other organic refuse mixed with lime and wood ashes and, after giving the bacteria sufficient time to complete their work, leaching out with water the potassium nitrate (variously known as saltpeter and niter). This created a competition between agricultural and political demand for the available nitrogen supply, with the constant threat of shortage. After the Spaniards conquered Peru, they made some sporadic attempts to utilize the Chilean nitrate for powder manufacture, but it was not satisfactory, since the compound found there is sodium nitrate, which takes up moisture from the air and makes the powder manufactured from it damp.

The idea of utilizing Chile nitrate for agricultural purposes seems never to have occurred to any one until 1809, when a German living in Bolivia pointed out that it would serve that need. Twenty years passed before the idea gained much ground, but a shipment of 110 tons was made to England in 1831, and sold for $108 per ton. Exports in 1840 were 8,600 tons and by 1860 had increased to 56,000 tons yearly.

After 1860, a new economic factor entered. First nitroglycerine and then smokeless powder were put on a practical basis, and for both of these sodium nitrate can be used, since the first step in the process is the production of nitric acid. So to the growing demand for fertilizer were added the requirements for these new uses and for other nitrated products, such as celluloid. The industry of nitrate production in Chile grew rapidly. The chart on p. 229 shows exports by years following 1880.

In spite of the rapid growth that arose from the fact that Chile had a complete monopoly of the world's supply of sodium nitrate, the industry was destined to be even more affected than the copper industry by developments that took place outside Chile, and in this case adversely affected. No country had been completely easy over a situation in which it was dependent for an essential material for military defense on another country and that a far distant one, and chemists were everywhere studying how to develop some alternative source of sup-

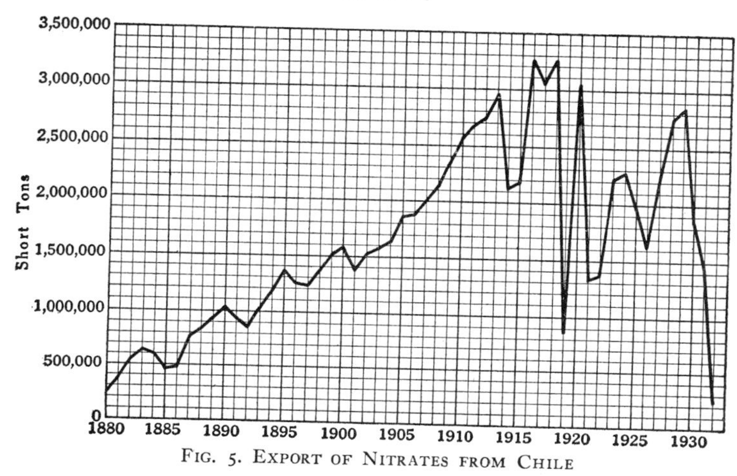

FIG. 5. EXPORT OF NITRATES FROM CHILE

ply of combined nitrogen. The export tax of about $12 per ton levied by the Chilean government furnished a commercial incentive for such attempts and toward the end of the nineteenth century a new and powerful incentive arose.

At the meeting of the British Association for the Advancement of Science, in 1898, Sir William Crookes, in his presidential address, prophesied that the deposits of nitrate in Chile would be exhausted by 1931. It is clear now that he was misinformed since when that year arrived a hundred years' supply was known, with large areas that may prove productive still untested. At the time, however, the statement created much concern, for the world was then in a period of general hysteria over an alleged impending shortage of various essential raw materials. Attempts to find alternative sources of supply of combined nitrogen were redoubled. Good progress was made, but no country had to depend on alternative sources until 1914, when in the World War Germany and

Austria were completely cut off from the Chilean supply, and even the Allies found it difficult to provide the necessary shipping for more than military needs, with a resulting increase in prices to agricultural consumers that was burdensome.

Every possible alternative source of supply was utilized, but the principal developments were in two fields. In converting coal into coke for iron blast-furnace purposes, about twenty-two pounds of ammonia per ton of raw coal can be recovered as a by-product if the proper means be utilized. Iron-makers had generally believed that by-product coke, as it was called, was not so satisfactory as that made by a process in which the ammonia was wasted, but by the end of the war it was generally recognized that this belief was erroneous and that by-product coke was as good or even better than beehive coke for making iron. In 1913, less than one-third of the coke in the United States was made by the by-product process, by 1930 over 90 per cent of it was so made, and the ammonia recovered amounted in 1929 to over 1,500 million pounds. It should be added, however, that a good deal of ammonia is also recovered from coal used in gas-making. It is evident that this source of supply is limited by demand for the coke which is the major product, and there is the objection to the use of ammonium sulphate for fertilizer purposes in that it "sours" the soil. However, the method is effective in meeting part of the demand for combined nitrogen that would otherwise have to be met by Chilean nitrate.

Even more important effects on the Chilean industry resulted from efforts to produce nitrogen compounds directly from the abundant supply of nitrogen in the air. The first attempts, in 1901, were along the lines of passing air through an electric arc, which causes some of the nitrogen to combine with the oxygen. The later improvements on this method can be briefly characterized by quoting the old story of the operation that was a perfect success, but the patient unfortunately died. The average price of Chilean nitrate in the United States in 1919 was 28.5 cents per pound, and in 1933 was 8.3 cents per pound. The power requirement for combined nitrogen production by the electric-arc process is so large, and the cost of production of hydro-electric power has so nearly reached an irreducible minimum, that it is now recognized that these processes have little commercial promise.

A second process that has had more commercial success is the production of cyanamid. Calcium carbide is made by heating coke and lime in an electric furnace, and nitrogen is then passed over the ground carbide at a high heat. The nitrogen is obtained from liquid air. Cyanamid has some chemical uses and is a good fertilizer, but has been more used for that purpose in Europe than in the United States. It contains free lime, which makes it somewhat objectionable to handle and may damage crops if used carelessly. It was estimated that it furnished about 12 per cent of the world's supply of combined nitrogen in 1928.

The most important process is the one known as the Haber, of which there are various modifications, but which consists essentially in causing nitrogen and hydrogen, under the influence of heat, pressure, and a catalyst, to combine directly to form ammonia. The main difficulty is the cost of producing and purifying hydrogen. A plant built at Oppau in Germany in 1913 to produce 10,000 tons of ammonia was expanded to produce 100,000 tons yearly during the war, and another plant, the Leuna, based on cheap lignite coal, was started and has since been expanded to have a capacity of 600,000 tons of fixed nitrogen yearly. There are altogether ten plants in the United States, the one at Hopewell, West Virginia, having a capacity of 177,000 tons of nitrogen per year. J. Enrique Zanetti estimated[9] that in 1932 there were eighty-three plants in the world, with a total capacity of 2,201,200 tons of nitrogen yearly, Germany alone having a capacity of 860,000 yearly. The ammonia thus produced is easily converted into sodium nitrate, which is even purer than the Chilean product.

These developments have, in thirty years, completely reversed the world situation as regards combined nitrogen. In 1900, the Chilean producers had a practical monopoly of the supply and the rest of the world was worrying lest the Chilean deposits would soon be used up, first sending the nitrate up to a prohibitive price level, and then making it unobtainable. Now it is evident that a permanent supply of combined nitrogen is available, and the Chilean producers are faced with the necessity of selling their product at a price lower than that of the other forms of fixed nitrogen if they wish to continue as producers. Our imports of Chilean nitrate into the United States which amounted to 617,000 tons in 1931 declined to 60,000 tons in 1932, and the price declined to little more than one-third the 1919 level.

This decline in Chilean imports, in 1932, was not simply a reaction to the fact that our capacity to produce synthetic and by-product nitrogen had increased to 600,000 tons yearly, whereas our normal peace-time consumption is less than 400,000 tons yearly, for our actual output that year was only 165,000 tons. Exchange rates were such that 350,000 tons of sulphate of ammonia were imported into the United States in 1932 and sold at

[9] J. Enrique Zanetti: "The Significance of Nitrogen," p. 48, New York, 1932.

prices below those prevailing here, while at the same time we exported 175,000 tons of sodium nitrate that had been produced here synthetically. Stocks of combined nitrogen here had increased about 30 per cent at the end of 1932 as compared to the beginning of the year. Domestic producers who had supposed that their production costs would be governed by the price of imported Chile nitrate found themselves, through the operation of exchange rates, in competition with producers of synthetic nitrogen abroad. An even greater potential threat to both the importers of Chilean nitrate and domestic producers of synthetic nitrogen is the plan of the Federal government, through the Tennessee Valley Authority, to engage in the production of fertilizer materials in order to promote agriculture in the South. What the outcome will be remains to be seen, but it would be possible for the price level of combined nitrogen to be set in this way at such a figure that Chilean nitrate could not be imported, nor could the existing plants for the domestic production of combined nitrogen operate at a profit. In short, the production of nitrate has turned into a keenly competitive business, with only the lowest cost producers likely to survive, and the strong possibility that the lowest-cost producers will be those most aided by government subvention.

The Chilean nitrate deposits, as already indicated, occur in the north desert area, extending altogether 450 miles north and south, and are scattered, some being as near as thirteen miles to the coast and others almost 100 miles inland. In general, they lie on the inner or eastern side of the coastal range of hills, and are connected by railroad with some ten ports, from Taltal in the south to Pisagua at the northern end of the region. The sodium nitrate occurs near the surface, seldom being deeper than twenty-five feet, and in beds that range from eight inches to fourteen feet in thickness. The sodium nitrate content is seldom over 40 per cent, and if it is less than 12 per cent cannot be worked at a profit, though grade and thickness are, of course, interdependent. Only about half the ground taken up for nitrate mining usually proves to be workable. Land belonging to the government is tested by government engineers and a lease, good till the land is exhausted, is auctioned off, the usual rate obtained being about six cents per Spanish quintal of the recoverable nitrate as estimated by the engineers. One-third of this amount is paid on receiving the lease, one-third in six months, and the remainder at the end of a year. There are no requirements for continuous production, nor is the actual output from the lease checked up; if the government engineers have underestimated the lessee gains, if they have overestimated he loses.

As the nitrate-bearing material is usually cemented into a fairly hard mass, it has to be broken before loading for transport to the extraction plant. Mostly this is done by hand labor aided by explosives, an advantage of hand work being that low-grade material can be sorted out and left behind. The workmen are paid by the cart-load produced, thus decreasing the amount of supervision. Steam-shovels and drag-line scrapers are used to a limited extent, but as, broadly speaking, these mechanical devices take everything, their use greatly increases the amount of material handled and transported; unless sorting is done before the material is put in the leaching vats its lower nitrate content requires a larger plant for the same output per day. Mechanical handling in the pits therefore mostly serves to decrease the total number of men required (and consequently the provision for housing and feeding them and looking after their welfare in a desert region), and as a check on progressive increases in the wage rate, since the higher wages go the more advantage there is in the use of mechanical equipment. With employment decreased through the shrinkage in demand for Chile nitrate the substitution of machines for hand labor has been checked.

The material as mined contains sodium chloride as well as sodium nitrate. Both are soluble in water, but the nitrate is much more soluble than the chloride, and to decrease the difficulty of later separating them the aim is to get out all the nitrate and leave as much of the chloride behind as is practicably possible. Two processes are employed, the older and more generally used one being the Shanks, which except on very rich material recovers little more than half the nitrate content. At two plants built by the Guggenheim interests,[10] in 1926 and in 1931, a different process is used which is especially adapted to large-scale operations with resulting lower unit costs. It differs from the Shanks process chiefly in using warm water rather than hot water for the leaching, and crystallizes out the nitrate at temperatures produced by refrigeration below instead of cooling down to atmospheric temperature. It is adapted to large-scale operations. The cooler solution breaks up the lumps less and does not yield such muddy solutions. The cooling water of the Diesel engines that furnish electric power for the plant is the principal source of heat. Interchangers to conserve the heat released in cooling the solution are utilized to the full, but a drawback of such highly efficient plants is the high cost of construction. The two Guggenheim plants, with an annual capacity of 1,300,000 tons of nitrate yearly, are said to have cost $70,000,000 to construct, or $54 per ton of annual capacity, and neither has ever operated to capacity. It is reported that the annual

[10] Anglo-Chilean Consolidated Nitrate Corporation.

operating deficit on the plant constructed in 1926 has reached $2,000,000 a year. In contrast it should be noted that H. F. Bain and H. S. Mulliken estimated[11] in 1924, that the average plant investment of the Shanks plants ranged from $20 to $32 per ton of annual capacity, and the probable operating profit represented 25 to 40 per cent on the investment.

There is an interesting general parallelism between the copper and nitrate industries in Chile. Both began in a rather small way and grew rapidly because there was a world market that would absorb the output at a price which afforded a good profit over the cost of production. There was never any market in Chile for the product of either. Capacity to produce was at first increased out of undistributed operating profits, the plants were not large, and the investment in them per unit of output was small. Eventually it appeared that by investing large amounts of capital, obtained outside of Chile, in huge, more highly efficient plants and processes, operating costs could be so lowered as to yield a favorable return on the investment. No fear seems to have been entertained lest the steady growth of world consumption should fail to persist, much less that it might show a decline. Not only did the last condition occur as a result of world depression, but its effects were heightened by the development of alternative sources of supply, on a large scale. Large deposits of copper capable of yielding the metal at a low cost were discovered in Rhodesia, while synthetic nitrogen plants capable of operating at a competitive cost, and with the advantage of being behind possible tariff barriers, were constructed in seventeen different countries. The two huge nitrate plants of the Guggenheim interests have

[11] The Cost of Chilean Nitrate, Trade Information Bulletin 170, United States Dept. of Commerce, Washington, D. C., 1924.

never operated to capacity. The plant of the Andes Copper Mining Co. attained 85 per cent of its theoretical capacity in 1929, but declined to 20 per cent in 1934 and may never attain the output it was designed and constructed to yield.

It should be noted that all these plants were technical successes, their processes worked successfully, and the operating costs were as low as expected. But a world-wide depression in consumption unfortunately coincided with the rise of additional sources of low-cost production. Whether such an eventuality should have been foreseen can be only a matter of opinion. But it should be noted that the effect on Chile of decreased sales abroad of two of its principal products would have been essentially the same whether foreign capital was invested in Chile or not, and the gradual passing of the small high-cost producer was inevitable, since it was brought about by the decline of production costs in other countries, and not by conditions in Chile. The outlook for the large investments of capital is not bright, but since it is capital derived from outside Chile, the effect of the loss on the country itself is minimized.

According to theoretical economics the plants with the lowest cost of production should continue to operate full scale in a period of reduced consumption and low prices, while all the others shut down. This has not been the case. Through an international agreement copper mines throughout the world were held down (with minor exceptions) to operating at 20 per cent of capacity, while in the nitrate field in Chile political and social necessities made it impossible to close down all of the smaller plants.

In spite of the fact that the war between Chile and Peru and Bolivia was provoked by the imposing by Bolivia of an export tax on nitrate produced by a Chilean company, Chile no sooner took possession of the region in 1879 than it also levied such a tax. The amount varied from time to time, but from 1897 to 1931 was about $12.38 per metric ton[12]; it was therefore independent of sales price. It commonly amounted to about one-third the usual selling price, and returned for a number of years $30,000,000 or more to the Chilean government in revenue. When prices declined it was natural for producers to clamor for a reduction in the export tax. Failing to obtain this, the Producers Association which had been organized to obtain concessions from the government turned to restriction of output as well as promotion of consumption, to maintain the margin between production cost and sales realization. Like all cartels for the control of production, difficulty was experienced in making it effective. Between 1884 and 1914, the industry was reorganized six times. Relations with the government became more close, the producers were granted exemption from import duty on bags, and were given reductions in railroad freight rates on nitrate, coal and petroleum. In 1919, the government formally approved the statutes of the Chilean Nitrate Producers Association, while in 1928 it created a nitrate bank backed by government funds, initiated the creation of a joint selling agency, and provided for propaganda for the promotion of consumption. Shortly afterward, it announced that a bonus would be given Chilean producers to countervail any reduction in price that might be agreed upon by German synthetic nitrogen producers. In 1925, the producers had offered to reduce the price by about 20 per cent if the government would reduce the export tax by a corresponding amount,

[12] The actual tax was twenty-eight English pence per metric quintal, and the amount in American dollars fluctuated with the sterling exchange rate.

but this offer was not then accepted. That decision now appears to have been an unfortunate one, because what the government gained by maintaining the rate was probably lost through reduction in the quantity of exports, while the maintaining of a high price level unquestionably stimulated the competition of synthetic nitrogen.

Exports of nitrates from Chile in 1910 amounted to over 2,700,000 tons, Germany and the United States each taking about one-quarter of the amount, which represented about 55 per cent of the world's combined nitrogen supply. Exports in 1923 were only two-thirds as much in quantity, and were only 32 per cent of the supply for that year. In 1926–1927, synthetic nitrogen was furnishing 80 per cent of the world's supply. By 1930, the situation in Chile seemed so desperate that a dramatic attempt at centralization and control was carried through.

The total number of firms then engaged in nitrate production was forty-five. The government and forty of these firms formed a corporation known as the Compañia de Salitre de Chile (more commonly abbreviated as Cosach) with $300,000,000 in common stock and $60,000,000 in preferred. Of the common stock, $180,000,000 was transferred to the government of Chile, which thus became a partner in the enterprise, in return for its agreement to abolish the export tax after 1933 and during 1931, 1932 and 1933 to take bonds of the company, up to $46,000,000 in amount, in lieu of the export tax. The Central Bank of Chile was authorized by law to carry these bonds as part of its gold reserve. Bonds were also issued in payment for some of the properties of the absorbed companies, and the Cosach assumed the outstanding bonds, amounting to $110,000,000 of the Anglo-Chilean, Lautaro, and other companies. So far as possible the properties were taken over for common or preferred stock.

The aim of this move was to permit closing down all the high-cost plants and concentrate production in five areas. The results are succinctly set forth in the following quotation.[13]

"When the Cosach was organized it was figured that the cost of nitrate at the plant would be about $6 a ton, which, with the addition of $10 for bagging, freight to port and ocean freight, would make the cost, delivered at consumer's port, $16 a ton. Assuming an annual production of 2,600,000 tons of nitrate, the overhead charges ($25,200,000) added $9.69 to the ton cost, making it $25.69. By selling the nitrate at $37.25 the company could show a profit of 10 per cent on the common stock. How far those assumptions have proved incorrect is shown by the fact that the nitrate production has fallen to about 1,700,000 tons in 1930–1931. The storage and interest charges on stored nitrate have further increased the losses of the company.

"The indebtedness of the company on May 29, 1932, was placed by the Chilean Minister of the Interior at what he termed the 'astronomical figure' of $158,302,562 and £21,656,342—part of the debt being contracted in United States dollars and part in British pounds. The interest and sinking fund charges on this debt amount to $32,000,000 a year and with the sales of nitrate at less than 1,000,000 tons for the year 1931–1932, the fixed charges would amount to over $32 per ton. Even if the original estimate of $16 per ton delivered at consumer's port without fixed charges be taken as approximately correct, the nitrate would have to sell for $48 a ton to cover

[13] J. E. Zanetti: "The Significance of Nitrogen," pp. 31–32, New York, 1932.

costs and at no profit. The 1932 price has been near the level of $31 a ton, until August, when the price was dropped in the American market to $21 a ton. The company is therefore heavily 'in the red,' and has defaulted on its bonds as well as on the bond of its subsidiaries, the Lautaro and Anglo-Chilean."

The aftermath of this bold but unsuccessful move was embarrassing to every one concerned, but particularly to the Chilean government. In recent years, the export tax had been furnishing about one-third of its cash revenues; in place of the cash it now had common stock which was not paying dividends and bonds which were not paying interest. It was forced to default on its own bonds, and eventually pushed off the gold standard. The plan to close down part of the plants involved an undertaking by the government to remove the workmen and their families to places where they could find employment. This has proved even more difficult than expected, on account of the general depression. As a result, the plants that had expected to operate but which on account of inability to sell the product should also have been closed down, were compelled by the government to continue operating to avoid increasing unemployment. Stocks of unsold nitrate were being piled up at the plants during 1933–1934, increasing the financial difficulties of Cosach.

In a situation such as this, the usual reorganization procedure would be for the bondholders to take over the property and for the holders of common stock to lose all they had put into it. But the Chilean government is the largest holder of common stock and the difficulties of its position are accentuated by the political pressure brought on the government by the former owners of the smaller properties who received only common stock for their holdings and in such a reorganization would receive nothing at all. It was proposed that a new corporation be formed with $132,000,000 in bonds and $65,000,000 in common stock, thus eliminating the preferred stock, reducing the common to about one-fifth of the original amount, and making the total liabilities less than half what they originally were. The original financial structure and the proposed reorganization are both so complex that it seems unnecessary to go into them in detail, but it may be pointed out that the chief original difficulty was that too high prices were paid for the various properties in taking them into the merger, creating a volume of debt which it is now evident could not be supported. The same pressure that originally created this condition is likely to prevent agreement to such a drastic reorganization as will permit profitable operation of the merged industry. The situation was made more difficult by there being so many different national interests involved, making it a diplomatic as well as a financial problem, while the frequent changes in the personnel of the Chilean government also made it difficult to complete negotiations.

At the time Bain and Mulliken studied the industry in 1924, the average cost of Chilean nitrate delivered at United States ports was about $35 per ton, and as it was then selling at about $50 per ton there was an excellent margin of profit. In the organization of the Cosach it was anticipated that costs would be lowered $8 per ton, making them about $27 delivered in the United States. But in 1932, the price at United States ports went as low as $21 per ton. Whether any reorganization of Cosach[14] that will permit operation at a profit can be made, therefore, depends not only on the conditions discussed above, but what level is finally attained in world prices for com-

[14] At the time of writing, July 1934, Cosach has been officially dissolved.

bined nitrogen. All attempts to limit production and stabilize prices through a cartel that should include producers everywhere have so far failed, but no one can prophesy what may happen in the future. Two things seem assured: (1) that Chilean nitrate has permanently lost its place as a natural monopoly of an essential raw material, and (2) the Chilean government can no longer count upon the large revenues it so long derived from the export tax on nitrates. It may have to give them up entirely in order to retain some degree of employment in the nitrate regions, on the railways, and in the ports which serve them.

The production of minerals for export in a country where there is no important domestic consumption is frequently discussed as though it were a case of "exploiting" the natural wealth of the country for the benefit of foreigners. Such a view tacitly assumes that a time will come when there will be a market for these natural resources at home, and at that time higher prices will prevail. The history of the copper and nitrate industries of Chile shows how fallacious it is to make such a general assumption. From what has been said above it is clear that if the time arrives when there is any considerable market in Chile for combined nitrogen, those needs can then be supplied by imports of synthetic nitrogen at price levels far below those prevailing in the period 1880–1919. If the Chilean nitrate industry had not been developed before 1930, it seems probable that it would never have come into being. Prior to 1930, it had yielded to the Chilean government something like $750,000,000 in direct revenue, which was available for promoting the economic welfare of the country, and in addition had paid out something like $1,000,000,000 in wages and salaries, mostly to Chilean citizens. It was a fortunate thing for Chile that its natural resource of sodium nitrate was early "exploited." If development had been delayed by a century it probably would never have occurred.

The case in regard to copper is generally similar, though not quite so clear. In the thirteen years following 1916, about $48,000,000 of American capital was invested in developing and bringing into production the copper deposit at Poterillos; if development had been delayed until 1930 there can be no doubt that the investment would not have been made. Whether the investments at Braden and Chuquicamata would have been made after 1930 no one can surely say, but it is probable the capital invested there would have been diverted to South Africa. If Chile had not had its copper and nitrate industries functioning during the nineteenth century and the first quarter of the twentieth, it seems probable that its general economic and social level in 1930 would have been on a par with that of a purely agricultural country such as Paraguay, instead of being what it is.

It may be claimed that it was very disturbing to Chile to receive for years a government revenue of $30,000,000 to $35,000,000 from export taxes on nitrate, and then to lose it. This is like arguing that a man who has always been poor is better off than a man who has had a large income but lost it. To any one who holds that view there seems nothing more to be said. No one would attempt to deny that the present depression in the nitrate and copper-producing regions creates difficult economic, social, and political problems in Chile, and that if those regions had never been developed the problems would not now exist. A country which remains on the basis of producing only what is required for the immediate subsistence of its inhabitants is almost immune to depression, though crop failures can produce conditions more severe than those of depression.

The plants engaged in the production of nitrate in 1879 had an annual capacity of about 17,000 metric tons and the capital invested in them was nearly 60 per cent Peruvian, a little less than 20 per cent Chilean, 14 per cent British, and 8 per cent German. By 1901, the annual production capacity had more than doubled, amounting to 36,500 metric tons, and the financial aspect was also greatly changed. Of the capital then employed 55 per cent was British, 15 per cent Chilean, 14 per cent German, 10 per cent Spanish and all other 6 per cent. Twenty years later, in 1921, the annual capacity had increased over forty times, the actual output that year being over 1,300,000 metric tons. Chilean capital then amounted to 51 per cent of the total, with British 34, Yugo-Slav 5, German 4, American 2½ and Spanish 2 in the plants then in operation.

It is easy to derive a somewhat wrong picture from the figures given above, for a considerable part of the capital reported above as Chilean belonged to naturalized foreigners, to Chilean-born children of foreigners, and possibly to corporations which had transferred their head offices to Chile to avoid taxation at home. There was indeed considerable growth in Chilean invested capital, but it functioned largely along with other factors mentioned in giving a Chilean aspect to existing enterprises rather than in starting new ones.

Iodine

Associated with nitrate production is that of iodine, of which Chile has for years been the principal world source. First discovered in 1813, this element is not only rather rare (ranking twenty-eighth of the elements in order of abundance), but relatively seldom occurs in concentrations that make its commercial production feasible. Seawater contains about a thousandth of 1 per cent, but seaweeds (kelp) hold a larger proportion. The ashes left from burning them contain about twenty to thirty pounds per ton, of which about twelve pounds is commercially recoverable. In Japan, Norway and Scotland some iodine, perhaps as much as 100 tons yearly, is still produced in this way.

The mother liquor resulting from the production of nitrate in Chile builds up from the 0.06 per cent average in caliche to a concentration of 0.3 per cent iodine in the form of sodium iodate. This relatively high concentration was early noted and a process for recovery of the iodine was devised. While it is used in small amounts for many purposes there is, unfortunately, no known large use and its recovery as a by-product in this way eventually so developed that the capacity to produce in Chile amounted to about ten times the world consumption, which was about 500 tons yearly. An Iodine Association was formed to control output and maintain prices; eventually sales agreements with all producers except those in Japan and Russia were obtained, and the price of iodine was pegged at $4.65 per pound, or over $9,000 per ton, a price which was satisfactory to the small producers outside the Association but not high enough to impel them to increase their output greatly. Control of production is aided by the government through the imposition of an export tax that is remitted on the quota allowed each producer. As the actual cost of production in Chile is about $1 per pound, this represented an additional profit of about $1.20 per ton on the nitrate of which it was a by-product. Since the prospect of greatly increasing consumption through a lowering of the price did not seem favorable, it was judged wisest to accept a large profit on the small amount that could be sold than a smaller profit on a hoped-for increase. This general situation obtained for some years, with continuing attempts to find new uses and enlarged markets for iodine.

The future of iodine production in Chile has recently been materially affected by developments which, as in the case of nitrate and copper, took place outside the country. Oil wells in California produce, along with crude petroleum, an emulsion of oil, salt water, and fine mud that has to be separated from the oil; it is desirable to recover the petroleum which this emulsion contains. In the course of research work on this problem a chemist, about 1927, noted that the salt water contained a detectable amount of iodine. The final outcome of the studies that resulted from this observation was that four years later there were three plants in California together capable of recovering about one-half ton of iodine daily, or sufficient to meet all the requirements in the United States. Apparently the cost of production of iodine from this source is such that the California producers can successfully claim the market in the United States for their own at a price well below the former pegged one. The price in 1931 dropped to $3 per pound and by the beginning of 1934 had further declined to about $2.25. This is probably less than the cost of production of some, or perhaps all of the iodine made from sea weed, and the lower price may lead to some increase in consumption when general industry improves. The present situation[15] is that the United States market, which amounts to about one-third of world consumption, has been permanently lost to producers of iodine in Chile unless they reduce

[15] P. F. Holstein: "Fortunes and Misfortunes in Iodine," *Chem. and Metallurg. Eng.*, vol. 39, p. 422, 1932.

their sale price below the cost of production in the United States. It is at least doubtful whether this would be profitable, since the gross realization from 70 per cent of the total at $3 per pound would be greater than that from 100 per cent of the total at $2 per pound, with a 30 per cent saving in production cost. At a time when it is already in financial difficulties, the Chilean nitrate industry seems unable to escape the loss of the major portion of the profit derived from the by-product production of iodine. Quantitatively, this amounts to a shrinkage to $1,000,000 in a profit that formerly amounted to about $4,000,000 yearly. It has been discussed in more detail than its importance may perhaps seem to deserve because it is a further illustration of the general fact that a mineral enterprise which produces for export and consequently depends on the world for its market is as likely, or even more likely, to be controlled by developments in world-production costs and methods than it is by fluctuations in total consumption.

Iron

The occurrence of iron ore has been reported from practically every province in Chile. In some places the iron is associated with copper ores, and some iron ore has probably been mined for use as flux in copper and silver smelters. It does not seem worth while to review the various attempts that have been made at local developments of iron-ore deposits in Chile, since the reasons why South American countries typically rely on imported iron and steel even when they possess workable deposits of iron ore have already been set forth in detail.

The most important iron-ore deposit in Chile is in the department of La Serena, province of Coquimbo, five miles from the port of Cruz Grande. Two hills are in large part composed of a mixture of hematite and magnetite so free from impurities that it averages 65 per cent iron. The deposit was originally taken up by two Chileans, who sold their rights to a French company which, with the aid of a government subsidy, built a blast furnace, Altos Hornos, at Corral, Valdivia. An attempt was made to use wood for fuel in this furnace, and between February, 1910, and April, 1911, 4,800 tons of pig iron were produced at a loss reported to have reached $300,000, or over $60 per ton. In 1913, the French company sold the rights to the mineral deposit to the Bethlehem Steel Co., which, after sufficient exploratory work to prove that the deposit contained over 100,000,000 tons of ore of good quality, low in silica, phosphorus and sulphur, proceeded to develop it for large-scale working. An electrically operated railway fifteen and one-half miles long connects the mine with the port at Cruz Grande, as the elevation of the deposits is so much above the port that this distance is required to afford a proper grade. The loaded cars descending generate electric power for hauling up the empty trains. At Cruz Grande a loading basin 900 feet long, 240 feet wide and forty feet deep has been excavated, and a steel dock which rises 123 feet above low tide provides two steel bins, of 15,000 tons capacity each, which discharge their contents by gravity into the holds of the ore-carrying vessels. The Bethlehem company operates a fleet of 20,000-ton ore-carrying vessels, but also charters tramp steamers when they can be obtained at favorable freight rates. This ore is delivered to the Bethlehem plant at Sparrows Point, near Baltimore, Maryland. So far as possible, the returning vessels take on a cargo of coal at the docks at Lamberts Point, near Norfolk, Virginia, which is nearly on the direct route between Baltimore and Cruz Grande, and discharge it at any port between where a market for it exists. The outstanding feature of this traffic is the provision of port facilities so that the time in port is minimized. A 20,000-ton cargo can be loaded at Cruz Grande in one and one-half to five hours, and a cargo of coal can be taken on at Lamberts Point in thirteen hours. While the Panama canal toll is 89.6 cents per ton on laden ships, this is computed by applying a standard factor to the cubic capacity of the ship, and for the dense iron ore amounts only to about forty cents per ton by weight. More recently some of the ships have been built to carry a back cargo of crude oil instead of coal.

The following table shows exports of iron ore from Chile, 1921–1932. The figures for 1921–1928 are those reported by the Chilean government, and are sometimes larger and sometimes smaller than the figures of mine production as reported by the company. The figures for 1929–1932 are those of mine production, not exports.

IRON ORE EXPORTS

	Metric Tons		Metric Tons
1921	48,393	1927	1,508,286
1922	222,947	1928	1,524,776
1923	679,187	1929	1,812,432
1924	1,049,860	1930	1,695,098
1925	1,234,094	1931	741,650
1926	1,396,406	1932	172,681

A detailed study of the iron-ore deposits of Chile, and of the general conditions applying to the possible development of an iron industry has been made[16] by Joseph Daniels, professor of mining and metallurgy at the University of Washington. He considers that there are seven other deposits beside the one described which might possibly yield ore for export.

[16] Joseph Daniels: "Mining and Metallurgy," pp. 200–206, May, 1926.

CHILE

Since 1925, various studies have been made in Chile as to the possibility of using the coke made from domestic coal as fuel supply for iron-blast furnaces in Chile. The usual conclusion of such reports as have been published is that it would be preferable to use electric furnaces, with charcoal as the reduction agent. It has already been pointed out, in the chapter on Brazil, that an electric furnace eliminates only about half the carbon requirement of a blast furnace; the other half, which is required for the reduction of iron ore to metal, must be furnished in the form of charcoal or coke. In 1920, the Chilean government financed another operating test of the blast-furnace plant at Corral and it is reported that an eight-day run produced 345 tons of iron at a cost of $25.30 per ton. Nothing has since been done with it nor, so far as can be learned, with an open-hearth steel plant, rolling-mill, and foundry which were also constructed.

COAL

Chile far outranks every other South American country as a coal producer, but that distinction is not a high one, since South America itself ranks lowest of the six continents in coal production, yielding only about one-sixth as much yearly as Australia and Africa (fourth and fifth respectively) and 1/250 of the normal annual output of North America. Australia's coal output is over two tons per capita per year, Chile's usual output less than one-third ton per capita per year. In quantity, the annual output is about the same as that of Italy, but per capita is about ten times greater.

The two principal reasons for this are the character of Chilean coal and the somewhat unfortunate situation of the deposits. The principal beds are in the provinces of Concepción, Arauco, and Bio-Bio, nearly 300 miles south of Santiago. Johannes Bruggen[17] has published detailed studies of these deposits and many other descriptions of their geology and operating conditions are available. J. del Fuenzalida described them, as well as the other coal deposits of Chile, in "Coal Resources of the World," published by the International Geological Congress in 1913. Fuenzalida estimated the coal reserves in Arauco as 1,872 million tons, and those of Concepción as 210 million tons, which are large amounts compared to the present rate of production. The total coal resources of Chile are estimated at 3 billion tons. Coal was discovered in Concepción bay as early as 1557 and the first coal mined in Chile was produced there in 1821, but real production began about 1840. The principal producers are the Compañia Minera y Industrial de Chile and the Compañia Carbonifera y de Fundicion Schwager. The first of these paid dividends of one peso in 1931, 1.60 in 1930 and 1.00 in 1932, while the second paid dividends of 0.75, 2.00, and 1.00 in the corresponding years. The beds dip at a low angle, about 15°, and extend out under the sea for more than a mile. As the overlying beds are not very tight and are broken by faults, the problem of handling water is a troublesome one, and some workings have been lost through flooding. There are several coal seams, but the companies mentioned work only the two thickest, four and four and one-half feet, respectively. At Lebú, in Arauco, there are seven seams, the thickest sometimes reaching six and one-half feet, but they are so folded and faulted the coal is badly shattered. The mines of the Colico district of Arauco lie farther inland, but have railroad service. Most of these

[17] Johannes Bruggen: "Informe sobre las exploraciones jeologicas de la rejion Carbonifera del Sur de Chile," Boletin de la Sociedad Nacional de Minera, 3rd Series, vol. 25, pp. 6–29, 49–84, 1913.

mines are also gaseous, and much attention to ventilation is required. The nearness of the coal seams to tidewater is not the advantage it might seem, for there are no modern port facilities, and the maximum rate of loading a vessel with slings and lighters is 200 tons per hour. The coal that is produced is used by the government railways, by coastwise steamships, for domestic use, for the manufacture of gas to a small extent, and for general industry. While it cokes it does not yield a coke of good quality for smelting purposes, being too friable. By mixing it with an equal quantity of imported coking coal a product of fair quality can be obtained. The coal is variously described as sub-bituminous and lignite, and as marketed has a heating value of 12,500 to 13,000 b.t.u. The ash content is 1½ to 7½ per cent; its high gas content causes it to burn with a long flame that creates difficulties in ordinary marine boilers.

Farther south, near Valdivia, there are also coal-bearing strata that are worked, but the production in September, 1933, was only 828 tons. In the extreme south, in the territory of Magellanes, there are various known outcrops of coal. In June, 1933, only two companies were operating there, the Rio Verde producing 2,634 tons for that month, and the Menendez Behety 2,431. The coal is a friable lignite, and breaks up so readily that for most uses it is necessary to use some binder.

North of Santiago there is no known coal, except in the province of Atacama, where at Ternera about sixty miles west of Caldera, material that is about 50 per cent ash has been mined desultorily since 1850, supplying small local needs. Production in January, 1931, was 584 tons. None was reported in 1933.

Chile's total coal production in September, 1933, was

reported as 124,163 tons, of which 67,055 were yielded by the Cia. Minera y Industrial and 39,998 by the Cia. Schwager. Two other companies at Concepción and one in Cañete together produced 11,842 tons for the month. Chile's total coal production in 1913 was 1,283,000 tons. Between 1920 and 1924, it ranged from 1,063,000 to 1,275,000 and in 1924 rose to a peak of 1,539,000 tons. In 1929, it was 1,507,866 tons, the production in intervening years being slightly smaller; in 1931, it was 1,107,000 metric tons. Imports in 1913 were 1,541,000 tons but in no year since 1920 have they exceeded 500,000 tons; coal, coke, and briquette imports were only 86,025 tons in 1928, and declined to 1,136 tons in 1930 and nine tons in 1931. The decline since 1925 can be attributed, in part at least, to an import tax of fifteen Chilean pesos per ton imposed on foreign coal in December of that year. Coking coal intended for mixing with Chilean coal is imported free of duty. To some degree the decline in imports is also a response to the substitution of fuel oil for coal in power and gas plants. A tax of three pesos per ton was imposed on imported petroleum in 1925, with provision for a gradual increase to twenty-one pesos per ton in 1935. This has not produced any marked effect on petroleum imports, which in the first years after 1925 tended slightly to increase, so far as gasoline is concerned. Fuel oil imports in 1925 amounted to 845,231 metric tons, and in 1926 were 797,101 metric tons. The 1930 figures show crude oil imports of 6,283,000 barrels, valued at 27,578,767 Chilean pesos and Diesel oil valued at 12,960,843 pesos. A ton of Chilean coal is equivalent in heat value to 4.4 barrels of fuel oil, so it is clear that although imports of foreign coal declined to a negligible figure (3,746 tons) in 1930, the crude oil imported had nearly equal heat value to all the Chilean coal produced in that year. It has been estimated[18] that imported oil displaced 1,650,950 tons of coal in 1923. In 1930, the liquid fuel imported was 500 times as much in value as solid fuel.

The decline in total fuel available for consumption after 1920 is evidently an index of declining activity in the nitrate fields and the decline in imports of coal between 1913 and 1920 also reflects the substitution of fuel and crude oil for imported coal. If, as seems possible, productive activity in the nitrate fields never regains its one-time peak level, the market for Chilean coal will probably continue curtailed. It would, of course, be easy to displace imported fuel oil by imposing a prohibitive tariff on it, but that would force users of fuel oil to provide coal-burning equipment and would increase their operating costs. The Chilean government is, therefore, likely to be in the dilemma of having to resist pressure to increase the tariff protection on domestic coal, in order not to handicap producers whose continued operation is as necessary to national well-being as that of the coal operators. One factor which favors petroleum imports is the fact that a tank steamer can discharge its cargo at the rate of 1,000 tons per hour, while in most Chilean ports coal can not be handled at a faster rate than 500 tons per day.

Further detailed study of Chilean production and consumption of solid fuel seems unnecessary here, for so many factors enter and must be taken into account to avoid being misled by too obvious interpretations put on figures. Thus, before 1925, imports of coke into Chile were about 35,000 tons yearly but in 1926 they almost doubled, reaching 62,000 tons. It seems probable that

[18] "Fuel and Power in Latin America," *Trade Promotion Service*, 126, United States Dept. of Commerce, p. 93.

this was not a response to increased requirements but, since a strike in Great Britain had cut off that country as a source of supply, consumers judged it wise to increase their stocks. The United States, Great Britain, and Australia are the principal sources of coke supply for Chile, and Great Britain usually sells about 10,000 tons of briquetted fuel there yearly. The only important consumers of coke are copper smelters.

Exports of coal from Chile are not large, and go only to adjacent territory. In 1924, exports amounted to 54,000 metric tons, but were only 15,000 the next year. In 1930, they rose to 122,757, valued at 2,358,735 pesos. Since Chilean coal needs tariff protection to hold its own in competition with imported fuel, it obviously cannot hope to compete in regions where the tariff protection is lacking, except in nearby regions, such as Bolivia and western Argentina.

Chile has large enough coal resources to meet all its fuel requirements for centuries. As a matter of fact, domestic fuel supplies only about half of domestic consumption, even with the assistance of tariff protection. The quality of Chilean coal is such that it cannot meet requirements for coke except by mixing it with imported coking coal, and even then the product only partly meets the needs of consumers. The coal is of fair quality for ordinary fuel purposes, but some consumers need high-grade or special coals for their special needs. The advantages of liquid fuel and the existence of a low-cost source of supply in California and Mexico will for some years at least furnish strong competition on a price basis at any reasonable tariff level. Chile's situation, therefore, is that it has sufficient coal so that it could meet most of its own solid fuel needs from domestic production, and it only buys foreign coal because the quality or the price makes it attractive. The tariff protection invoked about ten years ago has already practically excluded all but special grades of imported solid fuel from the Chilean market. With those markets restricted through lessened activity in nitrate and copper production, it seems probable that there will be increased pressure to give domestic coal more tariff protection against oil. A government commission appointed in 1923 reported that reduced domestic production had resulted from labor troubles, increasing cost of social services, electrification of railways, and loss of markets in the northern region because of inadequate port loading and discharging facilities and ocean freight rates that were proportionately too high for the domestic product to compete with on a price basis.

Blaming electrification of railways for decline of fuel consumption scarcely seems justified, since the total capacity of hydro-electric plants reported in 1928 was only 140,000 horse-power. The total waterpower resources of the country are estimated at 2,500,000 horse-power. The published figures of horse-power of prime movers in Chile are difficult to interpret; as nearly as can be determined from them, out of 87,000 horse-power of waterpower used in 1926, 70,000 was used in public utilities, food processing, paper-making and printing being the only other important uses. Even the public utilities derived more than two-thirds of their power from coal and oil. The region of abundant rainfall, in southern Chile, is one that is but little developed.

At the present time, the mining industry, the railroads, and public utilities are the only important consumers of fuel and power in Chile. The State Railways used almost half of the total domestic coal in 1930. Declining output of mineral industries produces a corresponding decline in railroad and port activities; so little mechanical power is used in general industry in Chile that even a considerable increase would not compensate for a slight decline in the larger fields of consumption. Gas produced from coal is extensively used and there is an annual production of around 50,000 tons of gas-house coke and some 5,000,000 liters of tar yearly, thus providing an important raw material for chemical manufacture. But chemical manufactures will be slow in developing in Chile. The fact remains that it and Colombia are the only South American countries that have sufficient coal resources to support a self-contained industrial civilization.

Petroleum

As yet there is no production of petroleum in Chile, and the indications that any can be developed are somewhat unfavorable, except possibly in the extreme south, where there is the least need for it. In 1927, the Chilean Congress passed a law providing that after December 26 of that year any concessions to petroleum lands not already developed should be canceled, and thereafter the production of petroleum should be nationalized. As there were no developed concessions, this automatically operated to nationalize all petroleum deposits. The Congress also passed a law (No. 4927) on January 5, 1931, reserving to the government the right to construct plants for the refining and hydrogenation of petroleum. A provision in the law that refineries already in operation could by registering obtain permission to continue operations and to double their output, was of only academic interest, since there was none of any importance. In the absence of a supply of crude there is no incentive for the construction of a refinery in Chile, except possibly a small one for the supply of local needs in the Valparaiso-

Santiago region, since the markets for refined products scattered throughout the length of the country could be reached with lower transportation costs by depending on direct imports. As mentioned above, there already is an import tariff on crude oil for the benefit of domestic coal production. To make it economical to bring in crude oil, refine it in Chile, and distribute it at high local transportation rates would require such a high tariff on imports of refined petroleum products that it seems improbable that the Congress would feel able to face the opposition that would be evoked from petroleum consumers. The nationalization of refining in 1931, therefore, evidences an intention to attempt to develop domestic refining through government subsidy rather than through invoking the aid of a tariff on refined products.[19]

The government made, in December, 1930, a contract with José A. Castillo to make geological studies and do prospecting for petroleum in five regions in the south of Chile on which he had made a preliminary report. This was a very favorable contract for the government since it provided that Sr. Castillo should meet the costs of the work and in the event of failure would not be reimbursed for them, whereas if they were successful the government would pay the costs and take over the wells. Nothing of any importance has so far resulted. A contract previously given to a Belgian company for drilling two wells near San Antonio likewise produced no result. A contract was also made with a German firm

[19] The *Boletin* of the Sociedad Nacional de Minera for October, November and December of 1931 printed the discussion in Congress on a national policy regarding petroleum which accompanied the passage of this law, but it does not seem necessary to review it in greater detail here. The President submitted to Congress on May 30, 1934, a bill modifying the code so as to allow concessions to Chilean corporations and to those having at least 60 per cent Chilean capital, with a majority of the directors Chileans.

in Santiago for geophysical exploration in the Magellanes region, and five men were set at work there, without productive result.

Attention has also been directed locally to the possibility of the production of petroleum products from shale. The general trend in all countries of the world after 1926 has been to regard the production of petroleum products from oil shales as not being within the bounds of present economic possibilities, except under some such arrangement as at Fushun, Manchuria, where the Japanese navy agrees to take the entire output at cost. It seems most probable that nothing will be done in Chile except in a small experimental way, and the country will continue in the future, as in the past, to rely upon imported petroleum products at prices set by international competition between the large marketing organizations. It is not necessary to repeat here what has already been said in the chapter on Argentina as to the difficulties met in attempting to develop a petroleum supply under government ownership and operation. The combination of unfavorable natural conditions and the restriction of petroleum development to governmental initiative seems fairly certain to insure that Chile will not have a domestic production of any importance at least for an indefinite period. While the question of the possible development of a domestic refining and distribution system hinges on a variety of economic conditions, it seems probable that the eventual outcome will be governed by political rather than economic conditions.

Other Metals

Lead production in Chile is ordinarily about 1,500 to 2,000 tons, which is chiefly exported in the form of ore or concentrates. The lead ores typically contain enough silver so that its recovery is worth while, and frequently also contain copper. Apparently, none of the producing properties is large enough to support a plant capable of producing commercial metal from the mixed ores, so they are exported.

Zinc ore and concentrate production amounting to 258 tons was reported in 1925 and increased to 2,360 tons in 1927, but again declined to ninety-six tons in 1929. No details are available and the fact that the London price of zinc in 1929 was only two-thirds that of 1925 may have been responsible for failure of zinc production to continue. There are no facilities for the treatment of zinc ores in Chile, and conditions there are unfavorable for their development.

From 3,000 to 10,000 tons of manganese ore have annually been exported from Chile in recent years. Production began about 1886 and reached a maximum of 50,871 tons exported in 1892. Four years later, exports dropped to 25,000 tons and by 1925 had declined to 1,300 tons. The reason for this was the competition in world markets for manganese ore from India, these being richer and cheaper to handle. The recent slight revival is apparently a reaction to improved shipping facilities to the ports and to the United States. The deposits have been described in detail by E. C. Harder;[20] they mostly lie inland from La Serena and Coquimbo, though one deposit lies about 120 miles farther north and there is another south of Santiago. Since there is no market in Chile for manganese as a raw material any production must necessarily be exported. And since in all the world markets for manganese ore, high-grade ore from Russia, Brazil and South Africa is keenly competitive, it does

[20] E. C. Harder: *Trans. American Inst. of Min. and Met. Engrs.*, vol. 56, pp. 62–68, 1916.

not seem probable that exports from Chile will greatly increase.

The only one of the minor metals of which an annual output is recorded is cobalt. Many silver mines in Atacama province have associated veins of cobalt ore, and similar ores are reported from Coquimbo and Santiago provinces. The most important deposit is about thirty miles from the port of Huasco, Atacama. In 1924, a production of thirty-four and one-half metric tons of 6 per cent cobalt ore was reported; in 1926, only six and one-half tons of 15 per cent ore, and in 1928 ten and one-half tons of 16 per cent ore. There is no bright future for these deposits, since the world's consumption of cobalt is small, and deposits at various places could produce larger amounts if there were a sufficient market for it.

Mercury ores have been worked in Chile since early times to furnish the metal required in gold- and silver-mining operations, principally in Atacama, Coquimbo and Aconcagua. The deposits are apparently not large enough nor rich enough to produce for export and there is no recent recorded production. Nickel ores are reported to have been produced in Atacama and Coquimbo, but there are no recent records of production. Antimony, bismuth, molybdenum and vanadium ores have been reported from various places in Chile but have not led to any recorded production. Whether they could be made the basis of a domestic supply when and if needed is not known. At present, and probably for many years to come, the domestic demand for the elements is slight, and only for the finished products of which they form a part.

Non-Metallic Minerals

Very little information regarding the production of non-metallic minerals, other than nitrate and borates, is available. Iodine and potassium chlorate have been discussed in connection with nitrate.

Northern Chile, as might perhaps be expected from its generally arid character and its active and extinct volcanoes, is one of the world's great reserve supply areas for boron minerals. Though boron forms one part in a hundred million of all living matter, it is rather a rare mineral element. It does not, however, present the appearance of rarity because where it is found is usually in easily available concentrations. The steam which issues from volcanoes contains boric acid, and lakes and lagoons in volcanic regions contain it in solution. When such waters evaporate or react with other minerals, they tend to form deposits of sodium borate (borax), sodium-calcium borate (ulexite) and a variety of other boron minerals. Volcanic regions that are arid are therefore likely to have workable deposits of boron minerals. Borax seems to have first been known in India and to have been used as a flux in soldering since very early times. Later, boron compounds came to have use in medicine and in the making of fusible glazes in ceramics. Boric acid has long been obtained from volcanic waters in Italy; borax is produced in California, Asia Minor, Tibet and India. Its use as a softener of water brought it to commercial prominence in the last quarter of the nineteenth century, and numerous borax-producing companies were formed in the United States.

The occurrence of ulexite in the nitrate-producing regions of Chile must have been noted soon after nitrate mining began, and its production was started in 1852. The high cost of transportation hampered production at first, but after the Antofagasta-La Paz railroad, which crosses the principal deposit, was constructed, production was stimulated and at one time reached the rate of over 40,000 tons yearly. This deposit is owned by Borax Consolidated, Ltd., of London, a £2,900,000 holding company that controls a dozen companies operating in various parts of the world.

In 1914, the United States was producing about one-half of the world's supply of crude borate minerals, Chile about one-third, Asia Minor about one-sixth, and Sicily, Argentina and Peru about one-twentieth. The attempts to produce potash in the United States during the Great War led to a greatly increased production of borax as a by-product. Although the consumption of boron compounds has increased greatly during the last quarter of a century, the production in the United States increased so much faster that our rather small pre-war exports grew to a record level of 68,000 tons in 1928. Production in Chile fell to 18,934 tons in 1928 and for 1929 no production was reported. Between 1926 and 1929, the carload price of borax in the United States declined over 40 per cent. Here again the course of mineral developments in Chile was controlled by developments in the United States. New uses continue to be found for boron minerals and it is possible that as world consumption increases Borax Consolidated, Ltd., which has larger interests in the United States than it has in Chile, may find it worth while to modernize its equipment in Chile to permit lower production costs there so that operations can be resumed at the price level set by the United States producers. More recent discoveries of potash deposits in the United States that are not associated with boron minerals may also change the situation, but about the best that can be hoped for the Chilean deposits is that they may some time regain their former importance. There is no domestic market for boron compounds in

Chile, nor is it likely that any considerable one will develop.

Lime and Cement

Lime and limestone amounting to 150,000 to 200,000 tons yearly is reported in the official government figures. Some of this is used as a flux in the copper smelters, some for the production of lime, and it is not clear from the figures whether that used in the production of cement is included in the latter figure or not. The "El Melon" cement plant began operation soon after 1900, but production figures first appear in the government reports in 1927, with an output that year of 96,000 tons, increasing to 111,000 in 1928 and 145,000 in 1929. Imports of cement amounted to 1,483,806 barrels, valued at $2,550,000 in 1930. There would seem to be no reasons other than purely commercial and political ones why Chile should not produce all its domestic requirements of cement with the help of sufficient tariff protection.

Little information regarding limestone in Chile is available. It is used for construction purposes, burning into lime, *etc.*, but there are no published data as to quantities. Curiously enough, a law (No. 4945) in February, 1931, authorized the President to expropriate deposits of carbonate of lime on the ground that it is needed for agricultural purposes. Quite possibly the real intent of this law is to make it possible for the government to control cement manufacture.

Clay production apparently increased from 3,000 tons in 1924 to almost 10,000 in 1929, but it is not known whether this represents an actual increase in production or is simply the result of reporting a larger proportion of the clay actually used, which in a region of adobe and brick dwellings must be a considerable amount. Of kaolin, 350 to 500 tons yearly are reported. There are no reported figures of stone production, except 1,000 tons of marble in 1929, although stone is used extensively in building and other forms of construction.

Gypsum production has been increased in recent years, being reported as 5,691 tons in 1924, 8,446 tons in 1925, 6,991 tons in 1926, 8,245 tons in 1927, 9,113 tons in 1928, and 15,434 tons in 1929.

Sulphur production has also been increasing, the reported figures, in tons, being as follows: 1924—9,765; 1925—9,072; 1926—8,928; 1927—12,500; 1928—15,670; 1929—16,300. The deposits have recently been described[21] in great detail by S. V. Griffith. Since the first important domestic use for sulphur was in making black powder for breaking the *caliche* of the nitrate deposits, it was natural that the sulphur deposits nearest to the market would be most exploited, but others occur all along the Andes, being associated with volcanoes, usually at such high altitudes and so remote from centers of population that there is but little incentive to work them. The methods used in producing pure sulphur from raw material that is 70 to 80 per cent sulphur are rather crude. Even with equally good methods the transportation costs alone would estop Chilean sulphur from competing in world markets with sulphur produced in the United States.

The Government and Mineral Enterprise

In Chile, the relations between the government and mining have been many and intimate, as is natural in a country where mining forms so large a part of the total industry.

In the nitrate fields the first relations were primarily

[21] S. V. Griffith: *Mining Magazine,* London, September, October, 1933.

political and resulted (a) in acquiring for Chile a large territory that had previously belonged to Peru and Bolivia, and (b) in providing the government, through a heavy export tax, with a large source of revenue. The need to increase this revenue led, through successive steps, into closer and closer business relations, until, in 1930, the government became a large stockholder in a nitrate monopoly.

Meanwhile, the trend toward socialization and nationalization of industrial enterprise had resulted in the formation, January 12, 1927, of a national mining bank, "Caja de Credito Minero." This was financed by authorizing it to issue 7 per cent bonds (of which the state guaranteed the interest), amounting to 40,000,000 pesos. These bonds were to be amortized at the rate of 1 per cent yearly. The law creating the bank evidently intended it to function as a financing agency, and it was authorized to make loans as large as 3,000,000 pesos each for the construction of smelters, mills, sulphuric-acid plants, the extension of existing plants and to provide working capital for existing plants. Loans of half that size might be made to mining enterprises as yet undeveloped, providing the owner could show that the ore reserves were such as to provide for amortization of the loans in twelve years. To qualify for loans, 75 per cent of the stock of the enterprise must be held by Chilean citizens, or by foreigners who had been resident in the country for more than five years. The loan would be secured by a first mortgage on the enterprise, construction was to be controlled by engineers representing the bank, and the bank retained the right to intervene in and superintend the operation of mines to which loans had been made. The borrower was required to pay 10 per cent on the loan annually; 7 per cent interest, 2 per cent commission, and

1 per cent amortization. Two other similar institutions, an agricultural bank and an industrial bank, were created at the same time.

In actual functioning, the bank created a subsidiary Caja de Fomento Carbonero, to deal specifically with the domestic coal industry. There seems to be no complete account of its dealings, but a loan of 300,000 pesos on April 21, 1931, to the Compañia de Gas de Antofagasta contingent on an agreement to use only domestic coal in its operation thereafter will serve as a general indication of the line of activities undertaken.

It can easily be understood that, after 1929, few operating companies were in search of capital to extend their operations, that any enterprise which was able to prove it could repay a loan in twelve years had passed the stage in which capital was most needed, and that entrepreneurs did not welcome the degree of control over their operations which was inseparable from the receipt of a loan.

Nevertheless, by January 1, 1930, the Caja Credito had loaned[22] 1,200,000 pesos for enlarging and modernizing a plant to produce sulphur; 2,000,000 pesos to modernize and reopen an old copper smelter at Guayacán; three loans, of 1,500,000, of 600,000, and of 405,000 pesos to different companies to construct cyanide plants; a fourth of 1,500,000 pesos to unwater a mine and construct a cyanide plant; two loans of 250,000 and 165,000 pesos respectively for the construction of flotation plants for copper ore; and a loan of 300,000 pesos to open a marble quarry near the Straits of Magellan.

In addition, the Caja was then proposing to finance, to the extent of 10,000,000 pesos, the building of a new

[22] F. Benitez: "The Mining Bank of Chile," *Eng. and Min. Jour.*, vol. 129, pp. 17–18, 1930.

copper smelter at Paipote, near Copiapó in Atacama, and to build three regional custom flotation plants for copper ore, at an approximate cost of 1,500,000 pesos each. One was erected at El Salado, near Chanaral; a second at Punta del Cobre, near Copiapó; and a third at Tambillos in Coquimbo. Constructed at a time when copper producers were trying to reach an international agreement to restrict the output of existing facilities to 20 per cent of capacity on account of the low price, the plants naturally found difficulty in securing custom ore to treat. At the same time, the Chilean government was wrestling with the problem of finding employment for the workers in the nitrate fields that arose when it entered into the agreement to form the Cosach (p. 239) and to concentrate production at the plants with lowest production costs. It was early pointed out that these men should be shifted if possible into the working of gold deposits, thus providing them with employment and the nation with a greatly-needed domestic supply of gold. The working of placer deposits was accordingly encouraged,[23] as involving the least preliminary expenditure of capital, but there were soon found to be limits to the possibilities in this direction. The regional concentration plants were therefore modified so as to treat gold ore, and according to a report[24] of the chief of the technical department of the Caja, the regional plant at Punta del Cobre had yielded 1,277 ounces of gold, that at El Salado 923 ounces, and the one at El Inca eighty-five ounces.

In 1931, the buying of gold ores was begun at Domeyko, and eventually a plant of thirty-five tons per day capacity was built. During the first eight months of

[23] A law passed in 1931 suspended for two years the requirement for a written petition for the right to work placer deposits.

[24] *Boletin Minero de Sociedad Nacional de Mineria,* p. 778, October, 1931.

1933 it treated 6,284 tons of ore with an average gold content of 17.42 grams, with an average recovery of 70 per cent. It was planned to increase the capacity to ninety tons per day and to improve its efficiency to 90 per cent recovery.

To facilitate the employment of nitrate miners in gold production, a Jefatura de Lavaderos de Oro was created. Between September, 1932, and March, 1933, the number of workers in placer-gold production was increased from 19,500 to 40,000, but by July, 1933, it had declined to 29,000. A report[25] of results in this field says that from the beginning (though the date is not given) till the end of July, 1933, a total of 1,396,097 grams of gold had been produced, of which 46 per cent had been yielded by "Lavaderos de Administracion," 23 per cent by concessionaires, 27 per cent by "particulares" (individual miners) and about 3 per cent by coöperatives. As the reported gross returns average about twenty Chilean pesos per worker per month, it can be surmised that gold-washing as a form of unemployment relief could not be considered more than a moderate success.

During the years 1929–1933, the general trend was toward increasing nationalization of mineral enterprise. Not only was petroleum production and refining nationalized, but various deposits under fiscal ownership were withdrawn from location under a former law. The general mining laws were revised. To indicate all the changes made would require too much space, and any one interested in details should consult the files of the Boletin Minero during 1931–1933, where not only the text of the laws and the regulations thereunder appear, but much discussion as well. It will be sufficient here to indicate

[25] *Boletin Minero de Sociedad Nacional de Mineria,* vol. 45, pp. 306–307, 1933.

that Chile was during this period developing a new viewpoint toward mineral enterprise, substituting for the original *laissez-faire* attitude the concept of undeveloped petroleum resources as a national asset, to be developed in the interest of the state, under state control and with state aid. Like our own Prohibition experiment, the underlying motive was admirable, but in both cases the essential difficulty was practical administration. It is too early yet to know what the results of so extensive government participation in mineral enterprise as Chile has undertaken will eventually be. It is natural for those who favor such enterprise to entertain high hopes, and equally natural for those experienced in the risks involved in mineral enterprise to fear that the final outcome will be a heavy burden of bad loans and of investments in plants which never will be able to amortize the investments.

The Caja de Credito Minero made a report of its condition on March 11, 1933. Like many other financial statements, this one is not easy to interpret, but it apparently shows loans and other advances amounting to over 186,000,000 pesos, investments in plants of 10½ million pesos, and an equal amount representing ore, concentrate, *etc.*, on hand.

The general situation of the mineral industry of Chile may be summarized by noting that its coal industry, for all practical purposes, consists of two moderate-sized companies which, with the aid of tariff protection, have been able to operate at a profit. For petroleum there is as yet neither domestic production nor refining, and whatever development takes place will be under national auspices. The iron industry is represented by one American-owned and operated mine which ships ore to the United States. All attempts to develop domestic pig iron and steel production have so far failed to succeed. The copper industry is characterized by large enterprises financed with foreign capital but is also represented by small enterprises, and the government has actively intervened to promote the development of small enterprises through technical and financial assistance.

The nitrate industry, on the other hand, developed into a monopolistic aggregation of capital in which the government became a partner. At the same time, the government is promoting individual enterprise in gold mining. The only discernible general policy is an apparent desire to promote mineral enterprise, the recourse to government subsidy and even operation springing from a recognition that where conditions are unfavorable private initiative based on the hope of profit cannot be counted upon for commercial development. This outward appearance of eclecticism is actually the outcome of the impact of two schools of thought in national affairs; one which believes in private initiative as the only reliable means of development, and a second which believes in nationalization of natural resources and government control and operation.

Whatever the outcome of the impact of these forces may be, it seems clear that Chile has come to the end of the period in which it can depend on the revenues derived from the export tax on nitrate to meet a large part of the expense of government, and it is doubtful whether the large interest which the government owns in the nitrate industry will in the future return even moderate sums. No other mineral commodity is in a position to bear heavy export taxes, and there are only two, copper and iron ore, which are even possibilities as revenue producers. The mineral industry has provided Chile with railroads, port facilities and employment for its citizens; to keep the citizens employed and the railroads and ports in operation requires production at operating costs that will permit the sale of the products in world markets. Government technical and financial aid will do something to secure that end, but to provide it involves a cost which may easily more than counterbalance such revenue as the mineral industry may be able to supply through direct taxation. It is clear that natural resources which await only appropriate development to produce great wealth do not exist, and that Chile's problem is that of developing a balanced economy in which domestic resources may wisely be developed provided the incurring of too heavy a burden of subsidy and direct aid is avoided.

CHAPTER X

PERU

SO MUCH of popular interest has been written about Peru that only brief references to its early history will be made here. The name is apparently derived from the Biru river, in Ecuador, where Pizarro landed on one of his early expeditions. Its present area of approximately 482,258 square miles, almost twice that of Texas, represents only a part of the region under Inca rule when Pizarro arrived in 1532. Huayna Capac, who ruled from 1482 to 1529, is reputed to have extended his sway from northwestern Argentina and northern Chile to Quito in Ecuador, and to have ruled over 10,000,000 people. Prescott's classic account of the conquest of Peru represents the Inca civilization as having arisen and fallen within a space of four centuries, but later archeological studies, which have been admirably summarized by Means,[1] indicate that it developed in the normal way from primitive beginnings over a much longer period.

As is true of the other countries along the west coast of South America, Peru is dominated by the Andes, with the usual eastern and western ranges and high plateau between and, as elsewhere, the principal metal resources come from the mountain territory. The Andes region was, in early geological times, a subsiding trough in which thick deposits of Silurian shale and Carboniferous

[1] P. A. Means: "Ancient Civilizations of the Andes," New York, 1931.

limestone were laid down.[2] The Mesozoic period is represented by a great thickness of limestone and sandstone. These strata were compressed into folds and broken by faults at the close of the upper Cretaceous age. The resulting mountains were worn down in part and then buried under a great accumulation of red mud and sand, and extensive flows of lava and volcanic ash. Again the region was compressed, old folds and faults revived, and the latest deposits sharply bent or broken. New mountains were uplifted and the deformed layers of sediments and lavas intruded by numerous small and large masses of porphyries and granitic rocks. The latter form great bodies, miles in their major dimensions. The copper, silver, lead, and zinc ores of the region are clearly associated with the smaller bodies of intrusive rock.

The great range of mountains thus produced had the same general course as the Andes, but otherwise probably bore little resemblance to them, since before the uplift which created the present range erosion had reduced the entire region to a country of subdued relief. The detritus accumulated in troughs or basins on the flanks of the range and is in part represented by the thick Tertiary deposits, from which the petroleum production on the north coast of Peru is obtained. Tertiary deposits everywhere flank the Andes, but are only a small part of the deposits that must have existed at one time. The present mountains result from a great, somewhat warped, uplift of the surface created by erosion of the Tertiary mountains. Its subdued forms can be seen throughout the plateau country at elevations where one would expect rugged peaks rather than smooth hills. Remnants

[2] The summary of the general geology of the country and the description of the Cerro de Pasco enterprise are based upon a paper presented to the Council on Foreign Relations by D. H. McLaughlin, formerly chief geologist to the latter company.

of the old surface occur on ridges between the new deep canyons, and the smooth skylines surprise the observer of distant views in the high country.

The uplift was undoubtedly accompanied by warping and faulting. The oil-bearing strata in the basin to the north are greatly broken and disturbed. The old surface itself is probably warped. Locally it terminates against faults that are still expressed as scarps in the present landscape. The uplift took place in stages, so that we have a succession of slopes on the sides of the deeper valleys, gentle in the upper parts, broad shoulders in the intermediate places, not too steep for cultivation and for villages, and precipitous cliffs at the bottom where the present torrents are at work with renewed activity.

The mineral deposits follow in distribution the rocks and processes to which they owe their origin. The zone of small intrusions of porphyries along the crest of the western range marks the axis of the belt of copper and related deposits. Gold appears to follow in part the older rocks on the eastern flanks. The more deeply dissected granite rocks on the west are generally barren. Petroleum is confined to the Tertiary basins on the flanks. Coal follows the belts of Cretaceous rocks in the high country. The geological history of the region gives the clue to the distribution of the mineral wealth, and only by gaining adequate knowledge of it can the search for new deposits be continued with reasonable hope for success. The obvious and easily found ores have been discovered, but there may still be hidden mineral deposits to be uncovered if suitable means are employed.

The three major geographical provinces of this region are a coastal desert, the Sierra, and the Montaña. The first of these includes plains and terraces along the coast, cut by transverse valleys with extensive sugar and cotton haciendas. The cities and towns are on irrigated lands or at adjacent ports. Desert conditions prevail as far up the western slope of the Andes as the level at which summer rains provide sufficient moisture for grass and crops.

The Sierra province extends across two lofty ranges and the intervening plateau, to forested slopes on the east. From the crest of the western range, Cordillera Occidental, there can be seen a bewildering complex of deep canyons and sharp ridges on its deeply dissected western flank, but to the east smooth, gentle, mature slopes prevail over great areas of 13,000 to 15,000 feet elevation, not far below the crest of the bounding ranges. In the central region the relief is mild, except where the old surface is dissected by the tremendous canyons of the eastern drainage, or where the ridges rise abruptly on the flanks. In southern Peru the roughness is smoothed out by a thick accumulation of volcanic rocks, capped by symmetrical lofty cones which give the range a totally different aspect. The eastern range, Cordillera Oriental, resembles the northern part of the western range, but is not as continuously lofty, being cut across by tributaries of the Amazon. It is not, therefore, so serious an obstacle to travel as the western wall.

The Montaña, a region still only partly explored, includes well-watered eastern slopes, from the sharp ridges and deep canyons of the intermediate zone down to the more subdued hills above the jungle. Travel is restricted to a few routes along difficult trails between canyons or upon navigable rivers.

Access to the Sierra from the west coast is through canyons, generally following rivers, with many detours across lateral ridges to avoid cliffs and bare rocks that baffled the abilities of native engineers. Modern routes, such as the railroads and few motor roads, likewise keep to the rivers, overcoming the steep gradients and precipitous walls by costly tunnels and high bridges. There were, in 1930, 2,481 miles of railroad in Peru, about half of it state roads owned and operated by the Peruvian Corporation of London. The two lines from the coast cross the western range at amazing heights; the Central Railway to Oroya climbs to a height of 15,680 feet in a distance of 106 miles from the sea at Callao.

It was such a land that Pizarro reached in 1532, after landing at San Miguel and marching overland to Caxamalca, where he met Atahualpa and his army. The capture of Atahualpa, his offer to ransom himself with a roomful of gold and two roomfuls of silver, and his subsequent execution need not be dwelt on here, except to make two points. The first is that the story, as usually told, creates an exaggerated impression of the amount of gold and silver which the natives possessed. The room was twelve by seventeen feet, and the Inca offered to fill it with gold to a depth of nine feet. Such a cube of solid gold would amount to some 33 million ounces, worth $660,000,000 at $20.67 per ounce. It was understood that it was to be filled with golden vessels, not ingots, and the usual estimate of the value of the gold collected for the purpose is $15,000,000. But even this seems to be based on assumptions of the exchange value of gold that may not be valid, and the most probable weight of gold indicated by the partition deed drawn up by the royal notary at Caxamarca in 1533 is less than 230,000 ounces, or not much more than a half of 1 per cent of the weight of the room *full* of gold. There are many stories to the effect that vast quantities of gold were thrown into lakes or otherwise concealed, as soon as treachery was suspected, but the real facts seem to be that all the gold and silver of the Inca state was in public possession, mostly in temples and royal residences, that the Spaniards got their hands on nearly all of it, and that the amount of it was not nearly so large as popularly supposed. Its concentration in a relatively few places gave it a spectacular quality which has been too strongly stressed.

The second point is that with the death of the Inca emperor his government crumbled and disappeared, like the dissolving of a lump of sugar in a cup of tea. Further resistance to the invaders was but slight, and though it has been estimated that there are still 3,000,000 pure-blooded people of the Inca race in South America there has been no native insurrection of any importance since 1786. Many consider that the elaborate organization of the Inca empire was its fatal weakness. Private citizens had neither weapons nor training in their use, and the complete government control of their activities had apparently robbed them of all initiative. Whatever the cause, the outcome was that people of a high indigenous culture fared no better than primitive aborigines under the rule of the invaders. The lot of both was that of slaves, and sixteenth-century South America much resembled Spain when it was under Roman rule, except that religious persecution was added to commercial exploitation.

Having collected and shipped away the visible stock of gold and silver, the Spaniards turned at once to the production of new metal. The general background of the Inca state made this relatively simple. A class of persons called yana-cuna, who, as field laborers, shepherds, and domestic servants were hereditary servitors, had long existed. The system of mita-cuna, under which citizens worked in the mines and elsewhere two-thirds of each year for the good of the state and priesthood, was extended by the Spaniards, a quota being fixed for each district, and lots being drawn to determine those who should work for a year at mining operations. Conditions at the mines were so difficult and the treatment of the laborers was so severe that drawing a mita lot came to be regarded as almost the equivalent of a death sentence. Men and women were required to marry early in order to keep up the labor supply by breeding.

So far as gold was concerned, the results were somewhat disappointing, chiefly because of exaggerated expectations. Both lode and placer deposits were found throughout Peru, but nowhere were they very large or very rich. With the abundant supply of what was practically slave labor it is estimated that 745,900 ounces of gold was produced between 1530 and 1600, or an average of about 11,000 ounces yearly. During the seventeenth century the yearly average rose to 16,000 ounces, and in the eighteenth, to 17,680 ounces. Between 1801 and 1810, it attained 25,000 ounces yearly, but thereafter declined, reaching the low level of 3,457 ounces yearly average for the five-year period 1891–1896. Between 1896 and 1926, production ranged between 30,000 and 120,000 ounces yearly of which an unknown fraction is a by-product of copper and silver mining. Of the 120,241 ounce total for all Peru reported in 1926, the Cerro de Pasco Copper Corporation contributed 48,864 ounces. In 1929 the operations of the Northern Peru Mining Co., described below, yielded nearly half the output, and the Cerro de Pasco about one-fifth.

Among the gold-mining enterprises that deserve mention are La Cotabambas Auraria Mina, a Peruvian enterprise in the Cochasayhuas district of Apurimac, for which a yearly output of about 20,000 ounces of gold and 80,-

000 ounces of silver is reported, and the New Chuquitambo Gold Mines, Ltd., organized in London in 1907 to reopen an old mine near Cerro de Pasco. This enterprise paid good dividends between 1907 and 1916, but has apparently since failed, as there are no recent reports on it. The most interesting is the Inca Mining and Development Co., on the western slope of the eastern range of the Andes, about 140 miles north of Tirapata, a station on the railway between Arequipa and Cuzco, and forty miles by pack trail from Huancarani. The elevation here is 5,750 feet as compared to 12,730 at Tirapata. The deposit was discovered in 1895 and taken over by Clarence Woods in 1928. The gold occurs in rich pockets in Silurian slate. A hydro-electric power plant and a 100-ton stamp mill were constructed, and, while no definite figures have been published, it is understood that the property quickly repaid the investment in it. In 1930, Mr. Woods obtained a concession to about 5,000 square miles in the valley of the Inambari river, which drains the region where the mine is situated. Mr. Woods claims that ten square miles of probable dredging ground exist. Airplane transportation is to be used to overcome some of the difficulties of the remote situation of the district.

Silver mining was much more productive. The Spaniards immediately began to work on an extended scale the deposits known and worked by the natives, and to explore for others. Adolf Soetbeer estimates that between 1601 and 1760 the average annual output was 3,250,000 ounces of silver yearly, Peru in 1661 taking from Bolivia the distinction of being the leading silver-producing country of the world, and retaining it till 1680, when Mexico became the leading producer.

Cerro de Pasco, now known as a copper mine, but

then a silver mine, was discovered in 1630. Only the oxidized ores were worked and up to 1911 are estimated to have yielded 300,000,000 ounces of silver. Until 1870, the rich ore was reduced in primitive smelters and the silver bullion transported by llamas 200 miles to Lima. Later reopened as a copper mine by what is now the principal mineral enterprise of Peru, it is described in more detail below. It would require too much space to review the history of the various silver-mining districts, with their periods of increase and decline, but it may be noted that silver production reached the high level of 4,864,000 ounces annually in 1801–1810, and its lowest level, 1,493,220 ounces annually, in 1881–1885. The production in 1896–1900 rose to 5,250,104 ounces yearly, in 1906–1910 to 8,845,904 ounces, and in 1916–1920 to 10,090,229 ounces yearly. The output in 1926 was 21,499,798 ounces, or 8.47 per cent of world output for that year. Peru, until recently, was the world's third largest silver producer.

Since silver production in Peru is now so closely associated with copper and lead production, it will probably be simplest to describe briefly the principal mining enterprises at the present time. The leading one is the Cerro de Pasco Copper Corporation, successor to the Cerro de Pasco Mining Co., organized in 1902 to take over 941 claims and 70,000 acres of land, of which 400 acres comprised about three-quarters of the Cerro de Pasco district mentioned above. Von Humboldt described these deposits in his "Essai Politique," and Miers in his "Travels in Chile" (1826) also describes them, though it is not clear that he personally visited the district. Miers' general impression was unfavorable, on account of the physical and political difficulties. At the time this company, in which the late J. B. Haggin[3] was the leading spirit, took them over there was but little known ore, or evidence on which estimates could be based. Nevertheless, the enterprise was started on what most mining engineers and geologists would have considered a highly speculative or almost fanciful estimate. The building of eighty miles of railroad, the finding and development of coal deposits, the construction of a smelter and the solution of many problems of unusual difficulty, because of the nature of the ores, the quality of fuel, the lack of skilled labor, the remote situation and the excessively high altitude, imposed huge difficulties, but all were eventually successfully met, and in a few years blister copper rich with silver and with some gold was on its way to the coast. The subsequent acquisition of a group of properties in the Morococha district, comparable in importance to the original Cerro de Pasco claims, and later the purchase of the mines of the Backus and Johnson Company at Casapalca and Morococha made possible the expansion of the original enterprise and the construction, in 1920, of the smelter at Oroya, to which ores are brought from all the mines for treatment. Copper continues to be the chief metal produced by the Corporation, but silver is still second in rank. Lead and zinc ores in important amounts have been developed in all three of the major districts and it is likely that in the future the output in these metals will be increased.

The production of the Corporation in round numbers in 1929 amounted to 97 million pounds of copper, 16 million ounces of silver, 40 million pounds of lead, and 20 million pounds of zinc. The copper and silver output

[3] His associates in the venture were H. K. Twombly, Phoebe Hearst, D. O. Mills, H. C. Frick, J. P. Morgan, W. D. Sloane, and S. W. Vanderbilt.

is shipped in the form of blister copper, the lead, with some silver, as lead bullion and concentrates, and the zinc largely in the form of sulphide concentrates.

The operations of the Cerro de Pasco Copper Corporation have become a dominant factor in the province of Junín and adjacent regions, and are an important influence in the financial and economic welfare of the entire country. The opportunity for work at the mines and plants has provided a better livelihood than formerly was possible in the Sierra region. A population trained to handle tools, to perform skilled labor, and to function as part of a smoothly running organization has been developed, and, for better or worse, is now an integral part of the life of the high country.

The Northern Peru Mining & Smelting Co., a subsidiary of the American Smelting & Refining Co., is the second most important group of mineral enterprises in Peru. In 1924 it purchased the copper-gold-silver deposits, about 70 miles east of Trujillo, of the Sociedad Minera Quiruvilca, and has equipped the property for an output of 20,000 tons of ore monthly. This ore is delivered by a 2-mile aerial tramway to the mill and smelter at Shorey, where the fine ore is concentrated by flotation, sintered, and smelted with the lump ore. The fuel for the smelter is derived from an anthracite mine at Callacuyan, four miles from Quiruvilca, with which it is connected by an aerial tramway.

This company also operates gold-silver mines in the Milluachaqui district, about 40 miles east of Trujillo. Some of the high-silica ore is sent direct to the smelter and the remainder is shipped by aerial tramway to a concentrating mill, of 200 tons daily capacity, which is about 7 miles west, at Samne on the Moche river. The mill concentrate is transported by motor-truck from the mill to Quirihuac. These properties are operated under lease, or a partnership agreement, as is also the Pataz gold mine, about 125 miles east of Trujillo, where operations began in 1924. The latter property is equipped with a hydro-electric power plant and a cyanide mill. About 3500 tons of ore is produced monthly, the cyanide precipitates being shipped by mule-back to Quiruvilca and thence by tramway and motor-truck to Quirihuac. The Shorey smelter is connected by aerial tramways to Milluachaqui and thence to Samne, on the Moche river, a distance of about 25 miles. A motor road 25 miles long connects Samne with the railroad of the Peruvian Corporation, Ltd., at Quirihuac. This group is the principal gold producer of Peru, and is second only to the Cerro de Pasco in silver production.

The next largest silver producer in Peru is the Casapalca mine of the Soc. Minera Backus y Johnston del Peru, now a subsidiary of the Cerro de Pasco Copper Corporation. The fourth most important district is that of Colquijirca where the Negociacion Minera Fernandini is mining a replacement deposit in limestone. The mine is especially interesting to geologists for the fine specimens of native silver that were obtained from the oxidized portion of the deposit. Most of the production has come from a rich shoot that contained 100 ounces of silver per ton. In depth, the ore becomes higher in copper and lower in silver. It is sent to the Cerro de Pasco smelter for treatment.

The Anglo-French Ticapampa Silver Mining Co., Ltd., bought, in 1903, the Ticapampa silver mine at the place of that name in the province of Huaraz, department of Ancachs. With a capital of £200,000, the property, which includes a thirty-ton lixiviation plant, a 140-horsepower waterpower plant and a sixty-mile railroad to the coast, paid a dividend of 6 per cent in 1905, increasing to 27 per cent in 1917 and 23 per cent in 1919, but has paid none since. Its silver production in 1929 was 10,820 kilograms, in 1930 was 8,222 kilograms, in 1931, 6,724 kilograms, and in 1932, 12,022 kilograms.

The Lampa Mining Co., Ltd., was organized in 1906 to acquire copper mines near Santa Lucia, in the province of Lampa. In 1913, it took over silver mines at Berenguela and a smelting plant at La Fundición, through an exchange of shares, and has also acquired sulphur properties. With a capital of £100,000, it paid a 7½ per cent dividend in 1928–1929, but has paid none since, and the company report for the year ended June 30, 1933, showed an operating profit of £51.

Ferrobamba, Ltd., a £150,000 British corporation, was formed in 1909 to acquire ten groups of claims, 2,255 acres, in central Peru, about forty-five miles west of Cuzco. Exploratory drilling revealed the presence of 6,000,000 tons of ore averaging 3.7 per cent copper. Further work was postponed until railroad transportation became available. In 1919, an option on the property was taken by Ventures, Ltd., a successful Canadian corporation, but, the general status of the copper industry being what it is, early development is not likely.

Many other small mining enterprises in Peru might be mentioned but, as the table below indicates, practically all the lead and zinc now produced in Peru is by the Cerro de Pasco Corporation, and it also furnishes much the largest part of the copper, silver and gold.

PRODUCTION IN 1929

	Gold, Ounces	Silver, Ounces	Copper, Pounds	Lead, Pounds	Zinc
Cerro de Pasco...	33,671	15,814,000	99,986,000	39,561,000	19,716,000
Peru, total.......	122,125	21,498,000	118,716,000	42,786,000	27,384,700

The lead produced by the Cerro de Pasco was shipped in the form of concentrate to the Selby smelter at San Francisco for reduction, while the zinc concentrate was shipped to Belgium.

Peru has the distinction of possessing not only the largest vanadium mine in the world, but one that is unique. At Minasragra, Cerro de Pasco province, about eighteen miles west of the Huaraucaca smelter, and at an elevation of 16,500 feet, there is a hydro-carbon deposit that had frequently been fruitlessly investigated as a source of fuel. In 1905, the manager of the smelter, A. R. Patron, had a complete analysis made of it, revealing that it contained a new mineral, which was named patronite, a vanadium sulphide which contains about 20 per cent of the metal. The deposit is a lens-shaped body about 300 feet long and thirty feet thick, which has been estimated to contain 28,000,000 pounds of vanadium. The property, 2,372 acres, has since been acquired by the Vanadium Corporation of America. The material as mined is burned, leaving an ashy residue which contains 50 per cent V_2O_5. This is shipped to Bridgeville, Pennsylvania, for reduction to ferrovanadium in electric furnaces. Production is somewhat irregular, depending on stocks in this country and the state of the market. The vanadium content of the material shipped in 1926 was reported as 1,577,000 pounds, in 1927 as 167,500 pounds, and in 1929 as 1,985,400 pounds. None was shipped in 1931. This mine ordinarily furnishes about 80 per cent of the world's vanadium output. Nowhere else in the world, not even in the vicinity of this deposit, has a similar one been found.

The quicksilver mines at Huancavelica are only of historical interest since they have been shut down ever since the war of independence, and subsequent attempts to reopen them have failed. Discovered in 1556, they furnished the quicksilver required in the processes used for the recovery of silver from its ores, saving the transportation cost from Spain, though this advantage accrued to the Spanish government, which owned and operated the mine, and is estimated to have produced 55,000 tons of metal. Quicksilver is not required in the modern methods of treatment of silver ores. There is no present production of quicksilver in Peru, nor likely to be any, as world markets are controlled by a joint selling agency of the Spanish and Italian governments, which is able to supply more than the market will normally absorb.

Bismuth occurs associated with other ores in Peru. At the San Gregorio mine, not far from the Huaraucaca smelter, it occurs as a separate mineral and the deposit was worked for a time after 1905. Recently the Cerro de Pasco Corporation has perfected a process for recovering bismuth from the flue dust of its copper smelter, and as the quantity thus made available is more than sufficient to meet market requirements, it has led to the suspension of operations elsewhere in Peru and in Bolivia as well. Exports of bismuth were reported as 618,300 pounds in 1931 and 184,300 pounds in 1932.

There is a small but steady production of antimony in Peru, where deposits occur in the departments of Puno, Cajamarca, Huanuco, and Junín. The reported production in 1928 was 108 tons, but exports declined in 1929 to eighty-six tons, in 1930 to forty-seven, and in 1931 to twenty-four. There is no domestic market for antimony in Peru and, as has elsewhere been explained, world market conditions make it difficult to compete with low-cost producers elsewhere.

A small amount of white arsenic is produced by the Anglo-French Ticapampa Silver Mining Co. as a by-product. The world supply of arsenic is so excessively large that this is of only minor importance in supplying limited domestic needs for agricultural purposes.

Tungsten ores occur on both sides of the Pelagatos river, which forms part of the boundary between Pallasca and Santiago de Chuco provinces. No production has been reported since 1925, when five and one-half metric tons of concentrate were produced, though production in 1913–1917 ranged from 235 to 471 tons yearly. So far as is known these deposits and another in Angaraes province are not likely to prove of much importance.

Nickel ores occur in at least two provinces, but no production has been reported.

Molybdenite occurs in veins in granitic intrusions in Jauja province and some 80 per cent concentrate has, in the past, been produced. Exports for 1931 contained 4.6 tons of molybdenum. Here too world conditions make it seem unlikely that its production is likely to attain much importance.

Little information is available on iron deposits in Peru. Miller and Singewald[4] describe veins of hematite, reported to be three to sixteen feet wide and two and one-half miles long and containing 50 to 70 per cent iron, which occur at Calleycancha and Aija, in Ancachs. At Huacravilca, in Junín, there are outcrops of magnetite said to contain 68 per cent iron, and Dueñas has estimated that the deposit contains 64,000,000 tons. Outcrops of hematite occur around the town of Tambo Grande, in Piura, but are too high in silica to be of much importance. Several other deposits of magnetite and hematite have been reported, but so far as is known there have been no attempts to develop a local iron in-

[4] *Loc. cit.*, p. 461.

dustry in Peru and the physical situation of the deposits makes it impossible for them to produce ore for export. There are foundries at Lima, Callao, Arequipa, Chichajo, and Cuzco.

Peru is one of the few coal-producing countries of South America, and here again the Cerro de Pasco Corporation is the chief producer. No coal occurs in Carboniferous strata in Peru, but deposits of lower Cretaceous age are widely distributed and in places of commercial importance. The important mines in addition to that at Callacuyan, already mentioned, are those of the Cerro de Pasco Corporation at Goyllarisquisga and Quishuarcancha. At the former there are four workable seams, varying in thickness from three to sixty feet; at the later there is one seam ordinarily five to eight feet wide, but in places increasing to as much as fifty feet. The coal as mined contains 27 to 32 per cent ash, but by washing is reduced to an ash content which makes it possible to coke it and use the resulting coke in copper smelting and for steam purposes. Until 1922, Peru's coal requirements were about 400,000 tons yearly, of which 60,000 tons were imported and the rest domestic production. Coal imports in 1928 were 28,585 tons, and 36,881 tons in 1929; domestic production in the same years was 177,513 tons and 213,654 tons, respectively. This decline in the apparent coal requirements was due to the increasing use of fuel oil, particularly by the more important railways and industrial plants. Since houses are seldom heated in Peru, and wood and charcoal are generally used for cooking, the demand for coal would be comparatively limited were it not for the inescapable requirement of coke for the smelters. The quality of coke produced would be unsatisfactory for iron production, though it can be used for copper smelting. Peru can not produce coal for export, and the future of the domestic industry is therefore chiefly linked with copper smelting. It seems doubtful whether coal production in Peru will

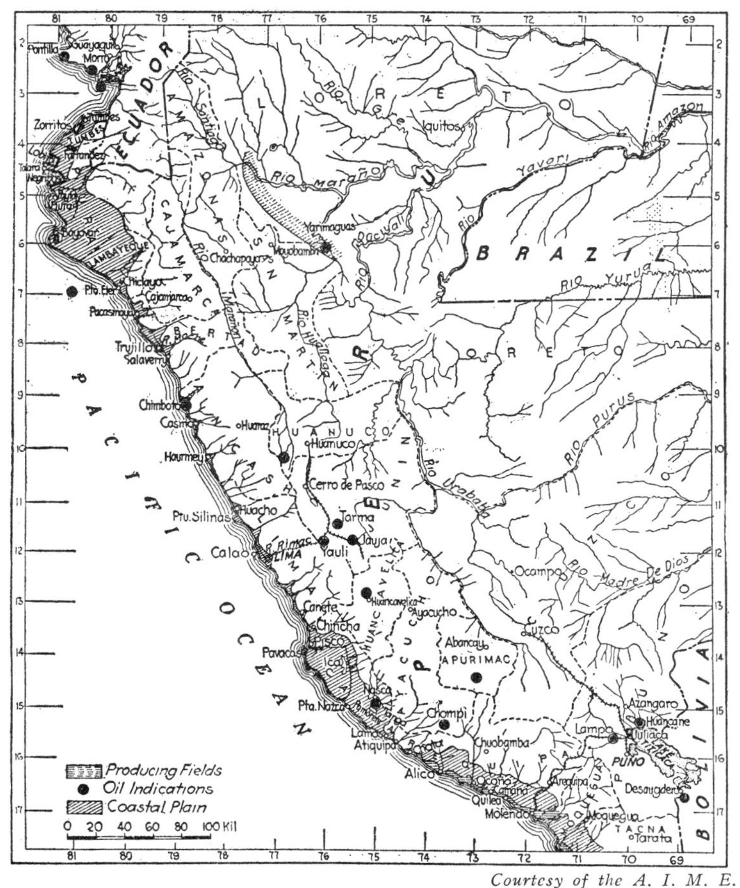

Courtesy of the A. I. M. E.

FIG. 6. PETROLEUM FIELDS, PERU

soon, or ever, exceed the 378,000-ton level which was attained in 1920.

There are three general areas in Peru which are either

productive of petroleum, or seem favorable for its development. A northern coastal area, extending from Tumbez south to Lambayeque, includes the three producing fields, Negritos, Lobitos, and Zorritos. The last of these is the oldest, twenty-nine wells having been drilled there between 1872 and 1898. Prior to 1900 the production of Peru was relatively insignificant (89,000 barrels in 1899), but since then has steadily increased, the output being controlled by market conditions rather than capacity to produce.

The principal producer in the Negritos field is the International Petroleum Co. which, to the end of 1931, had drilled 2,348 wells, of which 1,801 were still productive. It had produced to that time about a hundred million barrels of petroleum, or 80 per cent of the total production of Peru. The crude has an average gravity of 37° to 38° Be. Of the crude oil produced in 1930 about half was refined at the company's refinery at Talara, a little over a quarter shipped to Argentina, and the remainder to Canada, the United States and Europe. The property is excellently handled, about half of the gas or all that is not needed for fuel requirements being returned to the sands for repressuring. The company produces 30,000,000 gallons of casinghead gasoline yearly from the plants handling the gas. At the end of 1930, wells capable of producing 40,000 barrels were shut in. In addition to the 13,500 acres of productive oil land which this company controls on the La Brea-Parinas estate, near Talara, it owns extensive concessions in the area between Tumbez and Sechura on which it has done some exploratory drilling. Its production in 1933 was 11,205,362 barrels, nearly 85 per cent of the total for Peru.

The Lobitos Oilfields, Ltd., an English company, began producing in 1905. It has two fields, one at Lobitos and one twelve miles farther north, and controls a total of 300,000 acres, extending from the northern boundary of the La Brea-Parinas estate to the Mancora valley, and from the coast to the foot of the mountains. Its production, which was 1,700,432 barrels in 1925, rose to 2,627,988 in 1930, but has since slightly declined.

The field at Zorritos is owned by a Peruvian company, Establecimiento Industrial de Petroleo de Zorritos, which has maintained a small but steady production since 1884. In 1923, it had thirty productive wells and made an output of 93,343 barrels. Its output in 1930 was just under 54,000 barrels and in 1932 was 51,257. The company has drilled 343 wells, ranging from 300 to 700 feet in depth.

These are the only regions now productive in Peru, but in the Pirín field, about twenty-eight miles from Puno, on Lake Titicaca, about 50,000 barrels were produced annually between 1908 and 1910. This field is not regarded as being of any importance.

A region of great possible future importance is the Oriental, lying along the eastern foot of the Andes for 900 miles. Extensive geological investigations have been made there but little has been published as a result, other than that it contains a number of structures and some oil seeps. At present it is most easily reached by ascending the Amazon. In 1930, W. R. Davis and associates obtained a concession for building a railroad from Bayovar, on the coast, some 700 miles across the Andes to Yurimaguas on the Huallaga river. This carried with it a grant of petroleum and mineral rights to 2,500,000 acres of land on the Huallaga, Ucayali, and Marañon rivers.

In 1922, a new petroleum law was passed in Peru opening for leasing all petroleum lands in the Republic.

Production, which that year was 5,315,271 barrels, increased in 1928 to 10,750,676 barrels. By 1929 it had reached 13,422,000 barrels, declined to 9,899,749 barrels in 1932, but was 13,259,900 barrels in 1933. Until 1931, Peru ranked ninth as a world producer of petroleum, but in that year its output was exceeded by Argentina, and only slightly surpassed that of Trinidad. Its output could undoubtedly be considerably increased if market conditions warranted.

Only a few other non-metallic mineral commodities appear on the list of mineral products of Peru. Limestone amounting to from 60,000 to 95,000 tons, and dressed stone ranging from 10,000 to 20,000 tons is produced yearly, an output that is a measure of domestic needs. The lime reported ranges from 3,500 to 5,500 tons yearly. Salt production ranges from 28,000 to 32,000 tons per year. The latter is a government monopoly, conducted by the Compañia Salinera Nacional. Part of it is produced from saline marshes and lakes along the coast, while rock salt is mined at Playa Chica and at San Blas, in Junín.

Borates occur at the Laguna de Salinas, not far from Arequipa, where Borax Consolidated, Ltd., at one time operated. The general subject of borates has already been discussed, and as they do not appear in recent export statistics for Peru it may be assumed that the cost of production there is too high to permit competition in world trade.

Cement production in Peru increased from 11,278 tons in 1925 to 49,137 tons in 1929, but the one plant near Lima only partly meets domestic requirements, as 337,877 barrels were imported in 1929.

Iron and steel are the principal mineral commodities imported into Peru. Imports in 1929 amounted to 57,985 tons, valued at over twenty times the 36,881 tons of coal imported the same year. The figures given include rails and pipe, but not machinery and other manufactures principally composed of iron. Over 7,000,000 pounds of dynamite and similar explosives were imported in 1929.

Minerals constituted 67 per cent of the export trade of Peru in 1929. Gasoline, kerosene, fuel oil, and crude petroleum were the principal items, their total being valued at $51,304,000. Gasoline amounted to more than half of this, and crude petroleum most of the rest, the kerosene and fuel oil produced at the refinery at Talara going mostly into domestic consumption. The total of ores and metals exported in 1929 was $37,857,000 in value, more than two-thirds of it in the form of copper bars, with included gold and silver. Lead bars, with included gold and silver, amounted to $6,094,000 and silver ores and concentrate, $2,395,000. Zinc concentrates to a value of $1,870,000 were exported, and vanadium minerals valued at $552,000. In 1931, no vanadium was exported and the total of ores and products declined to $12,913,000, while petroleum and its products decreased to $14,847,000. Mineral commodities in 1931 amounted to only 50 per cent of exports, having been more affected by depression than agricultural commodities. Bullion and specie, which amounted to only $500,000 in 1929, rose to $6,261,000 in 1931.

These figures sufficiently indicate how dependent on mineral commodities is the foreign trade of Peru. It should also be noticed that over 80 per cent of the petroleum is produced by the International Petroleum Co., Ltd., a Canadian corporation in which there is a large United States interest. The Cerro de Pasco Corporation is almost equally dominant in the production of metals and ores. When the latter began to interest itself in

Peru, about 1900, the production of silver was less than one-fifth of that in 1929, and of copper, only a third. To bring about the development that has taken place required the investment of capital. This capital was not derived from the previous operation of Peruvian properties, it came from the United States. The first estimates for the development of the property called for $5,000,000, but unexpected difficulties and disheartening delays resulted in the investment by 1912 reaching $25,000,000. The "Copper Handbook" for that year (ten years after organization) said "in all likelihood, the production is now yielding a profit, with prospects of increased production and much lower costs ultimately." The Cerro de Pasco Copper Corporation, which took over the Cerro de Pasco Mining Co. in 1915, paid its first dividend in March, 1917. The operations of the Corporation have since been profitable, but it should not be overlooked that for a decade and a half they were unremunerative. Instead of the "exploitation" of a Peruvian resource, the development of this huge enterprise represents the putting of $25,000,000 of capital drawn from outside Peru into a venture that to many seemed hopeless. Certainly none of those who had previously derived a profit from the working of the deposit would have put up such a sum, or even a small fraction of it, in the hope of future gain.

The Peruvian company which produces petroleum in the Zorritos field has been operating there since 1884. Its production in 1932 was about one-half of 1 per cent of that of International Petroleum, though twenty-five years earlier their operations were of approximately the same order of magnitude. Why one company grew rapidly and the other but slowly can be only a matter of opinion, but almost any one would suspect that conservatism in the investment of capital, through the drilling of few and shallow wells, was at least a possible explanation why the Peruvian enterprise has shown but little growth in a period when the petroleum production of Peru increased fifty times.

Whether or not it is socially desirable to have the general economy of Peru dependent upon the rate of activity of foreign corporations, it is nevertheless true that without the foreign capital which they brought into Peru those activities would not have been possible and no corresponding development would have come about. Peru's choice was not between a few large enterprises or many smaller ones, but between the large corporations and relatively no production at all.

In attempting to forecast the economic future of Peru, it is natural to couple it with Bolivia. Like that country, its population is four-fifths Indians or mestizos, and 60 per cent of the population is classed as inactive, consisting of children, the aged, invalids, the indigent, and persons without work. But, unlike Bolivia, its agricultural activities furnish a considerable surplus for export as well as a more abundant living. Manufacturing has begun to develop; there are textile factories for wool and cotton; tanneries, shoes factories, flour mills, vermicelli, biscuit, chocolate, candy, soap and candle factories. Wine- and beer-making and the production of alcoholic beverages is widespread, and there are a few plants making glass, matches, cigarettes, brooms, trunks, hats, paper, barrels, and furniture. Much of this activity is concentrated at Lima and a few of the more important provincial towns, but a start has at least been made toward a balanced economy. When and if mineral production ceases to be of outstanding importance in Peru it will have led to the development of a state with sufficient resources for an adequate standard of living. Peru will never develop to the stage in which it will be comparable with Pennsylvania; possibly the Peruvians do not want it to do so.

CHAPTER XI

BOLIVIA

BOLIVIA is the only South American country without access to the sea. In colonial days it was known as Alto Peru and was a part of the viceroyalty at Lima till 1776. It was then transferred to the viceregal jurisdiction of Buenos Aires until 1825, when it was established as an independent state. At that time its territory extended to the Pacific Ocean. The boundary line between Bolivia and Chile was fixed at 24° south latitude by the treaty of 1866; it extended from there to the Peruvian province of Arica. This territory was ceded to Chile after the war in 1879, as has already been described, and a valuable mineral region, approximately 70,180 square miles in area, as well as access to the sea, was lost.

The present area of Bolivia is estimated at 514,600 square miles, but the possession of a large area in the Chaco is disputed with Paraguay. The population is officially estimated (1931) as 3,051,739, of whom about 1,587,000 are Indians, 975,000 mixed races, and 430,000 whites. With about the same area as Peru, it has only half as many people and has only a quarter as much external commerce. Minerals furnished 96 per cent of total exports in 1929, tin corresponding to 75 per cent. Except for a little crude rubber, coca leaves, and hides, there were few other exports. It is natural to think of

Bolivia as lying on top of the Andes, but as a matter of fact nearly three-quarters of its area is a plain sloping down to the east from the mountains and extending to the low level of the Madeira and Mamoré river valleys in the north, and to the Paraguay in the south. Sparsely populated, mostly with Indians, and with little industry except cattle-raising, this large area contributes but 3 per cent to the foreign trade of Bolivia. To the west of it is a region of high ridges and deep valleys known as the Yungas; an agricultural region in its lowlands, and with mineral resources, mostly unexploited, in the highlands. Still west of this is the high plateau amounting to about one-fifth of Bolivia, which is best known.

This is a region of scanty population as well as of high elevation. La Paz, the largest city, has only 110,000 people; Oruro, Potosí, and Sucre have about 30,000 each. Potosí lies at an elevation of 13,600 feet, or nearly as high as the top of Pike's Peak; Oruro and La Paz are at about 12,000 feet elevation. Around Lake Titicaca there are regions with as many as 125 people per square mile, but the general average is between ten and thirty, with many tracts that are essentially empty.

A glance at the map reveals that all of Bolivia lies east of Peru, since at the boundary line between them the Andes, which have a northwest-southeast trend in Peru, turn south. Thus the eastern plain region of Peru is west of the mountain heights of Bolivia. Even the narrow Chilean coastal region, which cuts Bolivia off from the sea, is east of most of Peru. The plateau of Bolivia, bordered on the west with mountains which average 15,000 feet in elevation, is much like the similar one in Peru, but is much wider. The general geology is similar to that of Peru, which has already been described.

Practically all of the economically productive area of

Bolivia is a region having an elevation of over 10,000 feet. In New York state practically all the area that has a higher elevation than 1,200 feet is economically unproductive and has for years been losing population to the lower-lying parts of the state. In Bolivia it is the low-lying eastern area which is economically insignificant. The reason for the contrast is that lowland Bolivia is so cut off from access to trade routes that it has never developed, while the mineral wealth of the highland is so great that it has been developed in spite of unfavorable physical conditions. Raimonde enthusiastically described this highland as a "table of silver supported by columns of gold."

Invading the region from Peru in their search for gold and silver the Spaniards early discovered the valley of the Tipuani river, about sixty miles east of La Paz. Though accessible only over a pass 14,500 feet high, both placer deposits and lode mines in this general region were worked by drafted labor, as described under Peru, and between 1493 and 1800 Bolivia yielded[1] about 10 per cent of the total gold output of South America, ranking third for that period. Between 1901 and 1925, Bolivia ranked twelfth among the thirteen countries, only yielding about three-quarters as much as little Uruguay. It now ranks eleventh with a yield that since 1880 has seldom exceeded $60,000 yearly. There have been temporary increases at various times from attempts to work these deposits by modern methods, such as that of the Incahuara Dredging Co., on the Rio Guanay, and the hydraulic operations of the Bolivia Gold Exploration Co., on the Tipuani. Two great difficulties are obtaining

[1] Spanish reports claim that Bolivia produced $2,000,000,000 in gold between 1540 and 1750. A sounder estimate is $150,000,000 between 1493 and 1800.

laborers in so remote a region, and lack of supplies. The Bolivian company had to start farming in order to provide a cheap and abundant food supply at the mines. Since the gold content of the placers is rather high they provoke repeated attempts to overcome the absence of transport and their remote situation. A false report of rich discoveries in 1912 caused a rush that soon subsided. It is possible that the use of airplanes, recently introduced by the Bol-Inca Mining Co., operating on the Kaka river, may overcome some of the transportation handicaps and materially increase the gold output. Gold deposits also occur in southern Bolivia near the border of Argentina but they are apparently of minor interest.

Farther to the south the Spaniards found rich silver deposits, the richest in the world. The Cerro Rico of Potosí was discovered in 1544 and its veins were so rich that the town of Potosí soon had a population of 160,000. There are no precise figures as to the amount of silver obtained, but the claim that during the first forty years the yield was $70,000,000 worth, and the "royal fifth" amounted to as much as $600,000 per year, is probably somewhere near the truth. The Potosí mines were the richest, but there are altogether some fifteen silver-producing regions in Bolivia, of which Oruro, Porco and Huanchaca are perhaps the most famous. Between 1493 and 1600, Bolivia produced almost one-half of the world's output of silver. Peru ranked second as a world producer in this period and Mexico third, each contributing one-eighth. As the ores were somewhat different from those in the principal silver-producing regions of Europe (Austria and Germany) a good deal of metallurgical development took place, but even at their best the methods were rather crude.

After the war of independence silver production declined to about 1,000,000 ounces yearly, drafted labor being no longer available. It gradually increased until it reached 22,000,000 ounces in 1893, and then again declined to about 2,500,000 ounces per year from 1914 to 1920. Production has now again increased because of increase in tin output, silver and tin being associated in many of the deposits. Silver is now, however, distinctly secondary to tin in Bolivia and it will perhaps suffice here to discuss its tin deposits in detail. Readers interested in Bolivia's past silver production and those deposits not associated with tin should consult Miller and Singewald's book.[2]

The tin-silver-bismuth-tungsten ore deposits of Bolivia are of intense interest to geologists, since they illustrate what are termed a metallogenetic province and a metallogenetic epoch. In plain words, this means that the deposits of a large area are all of similar type and were formed at the same time. There is no close agreement as to details, but it is generally agreed that these minerals have been deposited in fissures in volcanic rocks at a relatively recent period, perhaps the Pliocene. They form veins that are often narrow and rich. Near the surface many of them were rich in silver, but in general tin is the most important mineral present. Nine districts are ordinarily described as tin rather than silver-tin districts; seven are classified as silver districts. The veins may be only an inch wide, or disappear entirely, or they may be a yard wide, contain 25 per cent tin and persist for a considerable distance. It can readily be understood that a large problem in mining is to locate ore bodies that are big enough and rich enough to be worked at a profit.

Tin ore must have been worked by the natives before

[2] Miller and Singewald: "Ore Deposits of South America," pp. 121–134.

the Spaniards came, since they produced bronze. In the Colonial period there was little interest in tin ore, since there is no fuel to reduce it to metal on the plateau and the concentrate could not be shipped to Europe in competition with Cornish tin ore, then available in sufficient supply to meet the rather limited requirements for tin in the world of that day, and even to meet the increased requirements when the tinning of iron and the manufacture of tinware came into use. In addition, the Cornish ores, and the later-discovered alluvial ores of Malaysia, were clean, pure oxide, so their reduction to metal was simple. The Bolivian ores are complex and the production of metal of good quality from them is much more difficult. Though tin deposits were worked at least as early as 1880, until 1895 the production of tin in Bolivia was unimportant. In 1896, the average price of a long ton of tin in London was £59½. It rose steadily until it reached £209½ in 1912, declined, then rose again to £329½ in 1918. The ten-year average price 1924–1933 was £211 4s. 8d. as compared to £94 13s. 8d. for 1894–1903.

This increasing price furnished an increasing incentive to work the Bolivian tin deposits and contributed to the rise of Simon I. Patiño, who is the Andrew Carnegie of the Bolivian tin industry. The story (or legend) of how he, shortly after 1900, accepted a mortgage on the Salvadora mine at Uncia in payment for a debt owed his employer in Oruro, and how his irate principal forced him to keep the mine and pay the debt himself is too well known to need repetition. Señor Patiño found the means to do this and to raise funds for exploratory work that was eventually rewarded by the discovery of a vein a foot wide and rich in tin. Out of the profits from working this he increased his holdings and eventually acquired extensive interests in banks and railroads, so that he is reputed to be the fourth richest man in the world.

All the tin deposits at Uncia are now a part of Patiño Mines and Enterprises, Cons., a $50,000,000 corporation organized in Delaware in 1924. In 1929, General Tin Industries (a subsidiary) purchased Williams Harvey & Co., Ltd., in which the National Lead Co. held a half interest that was later exchanged for Consolidated Tin Smelters. At an earlier time ownership was divided between the Empresa Minera La Salvadora, the original Patiño company, and the Compañia Estanifera de Llallagua, a Chilean corporation which exploited the veins on the opposite side of the mountain. A feature of interest to a visitor to the property was the underground defenses, including a steel barricade, employed in the vigorous conflict between the former rival interests. The reason for this arose from the peculiarities of Bolivian mining law.

Mining rights to metalliferous veins in Bolivia are now based upon a claim of 100 meters square, for which an annual fee of four bolivianos must be paid. Any mineral which is found beneath the vertical projection of the surface area belongs to the claim owner. But in earlier days, especially at Potosí and Machamarca, the claim owner was privileged to extend his operations in any direction beneath the surface, provided only he did not cross the workings of any other operator. The difficulties and conflicts that would arise under such a system can be easily imagined, and the only final solution of them is the securing of complete control of any mining district. This may not be easy in the case where recalcitrant claim owners are present. There is no requirement for performance of work on a claim. If the annual fee is not paid the claim lapses. If a claim proves valuable it is easy to claim ancient titles and unlapsed rights, which lead to provoking delays in litigation, passing through the offices of notary, prefect, and procurator to final appeal to the government. In 1910, the government created a Comision Tecnica of engineers for the provinces of Pacajes and Bustillo to map the mines, delimit concessions, and attempt to reconcile conflicts. This led to much improvement, but it still remains true that it is necessary, before beginning mining operations in Bolivia, to devote much time and effort to insuring that all necessary legal safeguards have been employed.

Since the property and operations of Patiño Mines and Enterprises have recently been described in much detail by R. R. Beard,[3] it will be enough to state here that on its 4,700-acre property at Uncia there are over forty veins, cutting through a batholith of quartz-porphyry. The veins are usually narrow, averaging perhaps ten inches in width, and the most important one has been worked to a depth of 2,800 feet. Altogether, there are about eighty miles of workings, and some 3,000 workers are employed underground, under foreign supervision. The broken ore is sent to a modern mill which has a capacity of 2,000 tons daily. Formerly, high-grade shipping ore was sorted out by hand. Details as to the concentration practice, which now includes froth flotation, can be found in the paper cited.

The finished concentrate, called *barrilla*, and containing 62 per cent tin, or more, is dried and packed in paper-lined jute sacks for shipment to the tin smelter of Williams Harvey & Co. at Liverpool, where it is reduced to metal. About a third of the power used in

[3] R. R. Beard: *Eng. and Min. Jour.*, vol. 130, pp. 107–109, 169–173, 235–237, 1930.

the operations is derived from hydro-electric plants and the remainder from Diesel engines.

Though the Patiño tin properties are the most important in Bolivia, there are many others. Tin production in Bolivia in 1929 amounted to 42,991 long tons of contained metal, of which Patiño Enterprises produced 20,926 long tons, or a little less than half. Other producing companies are shown in the table below, which also gives the production quota assigned to each, by the International Tin Committee, for the six-months' period, July to December, 1932 and thus indicates their approximate relative importance. The actual allotments have since been increased as the general tonnage has been raised.

	Quota
Patiño group	4,194.75
Compañia Unificada de Potosí	885.00
Caracoles Tin Co. of Bolivia and Compagnie Aramayo de Mines en Bolivia	377.38
Compañia Minera de Oruro and Empresa Minera de Vinto	358.38
Compañia Estanifera Morococala	177.55
International Mining Co.	156.66
Fabulosa Mines Cons.	151.28
Trepp y Compañia	85.45
All others	1,074.54

Of these, the Caracoles Tin Co. is a $40,000,000 corporation formed in Bolivia in 1922 by Guggenheim Bros. of New York, now operated by the Cie Aramayo de Mines en Bolivie, a Swiss corporation owning the British Aramayo Francke Mines, Ltd. The Fabulosa Mines, Cons., is a Bolivian corporation in which a controlling interest is owned by the £500,000 Bolivian and General Tin Trust, of London.

It will be seen from the foregoing that tin-mining in Bolivia is to a considerable degree an international enterprise, a good deal of Chilean as well as European and American capital being involved but Bolivian capital now dominates the situation. The presence of foreign capital was an advantage in putting into effect an international cartel to stabilize the price of tin when, after 1927, its price began to fall. World capacity for the production of tin, stimulated by the high price, had increased so much faster than consumption that in 1930 the stocks in hand caused the price to decline to about half that of 1926–1927, a figure lower than any previous year since 1909. Bolivia's output in 1929 was 42,991 long tons as compared to 29,600 in 1926, but production in the Federated Malay States had increased to 67,040 tons from 45,946 tons in 1926. An international agreement was negotiated under which the governments of the tin-producing countries undertook to place a limit on the amount permitted to be exported. This agreement was made early in 1931 and provided for a 26½ per cent reduction in output compared with 1929, when world production was 186,874 long tons. A pool formed to buy metal and hold it off the market had in its hands early in 1932 nearly half the stocks of 51,000 tons, which had not materially decreased. Five additional reductions in the quota were made, bringing it down to 101,096 tons for 1932 and eventually to 54,000 tons for 1933. Actual world production was reduced to 95,871 tons in 1932 and to 87,485 tons in 1933. The price of standard tin in London, which had declined to £104 8s. per long ton in May, 1931, as compared to £302 13s. 10d. in April, 1927, by December, 1933 recovered to £227 14s. 10d. Late in 1934 it was decided to fix the quota for 1934 at 71,000 tons to prevent too much of an advance in price; still later the quota was increased to 80,000 tons. By April, 1934, the price had reached £243 and the cartel was using its pool to attempt to keep the price down instead of to advance it.

We are not concerned here with details of the operation of the international cartel; what is important is that the tin production of Bolivia will hereafter be a definite fraction of a world quota fixed by international agreement. The allocation of Bolivia's quota among the various producers is done by local agreement. The price level reached in 1926–1927 was one at which production by Bolivian mines was frequently unprofitable. The agreed production for Bolivia was 3,862 tons monthly in 1929, reduced three times in 1932 to 2,402 tons monthly, and three times in 1933, finally being fixed at 1,224 tons monthly, or less than a third of its 1929 output. The present price level represents an excellent margin of profit for producers, and they will undoubtedly prefer to maintain it rather than to increase output, which has been fixed at about 28,000 tons for 1934.

Several of the tin-silver deposits of Bolivia contain appreciable amounts of bismuth, especially those in the departments of La Paz and Potosí; The Aramayo Francke Mines, Ltd., early in this century commenced its production. Between 1904 and 1909, the average annual production was about 2,000 tons, valued at £93,965. As the only other important producer at that time was the Royal Mines in Saxony, an agreement was made to peg the price at £732 per ton. An export duty of £3 10s. was imposed by the Bolivian government. In 1929, the exports were 112,135 kilograms of metal and 38,937 kilograms in ore, concentrates, and residues; by 1931, it had dwindled to 26,581 kilograms and had almost disappeared in 1932. Unfortunately, there is no large demand for bismuth, as no important uses, such as require large amounts, have been found. The limited market falls into the hands of the lowest-cost producer and, as elsewhere indicated, that is likely to be the Cerro de

Pasco of Peru, though in Canada, Australia, Japan and Spain small production continues. At this instant it appears unlikely that bismuth will be of much future importance in Bolivia.

Tungsten ores occur in Bolivia, associated either with tin, or in separate veins in the tin regions. The important new uses for tungsten developed early in this century caused its price to advance and production began to develop in Bolivia. By 1913, it had reached 339 tons of concentrate containing 66 per cent tungstic acid and under the influence of war prices increased to 4,846 tons in 1917. By 1921, it again declined to 212 tons and in 1928 to twenty-nine tons, the main reason being that a source of supply had been discovered in China, which was able to supply world requirements at a price level which other producers could not attain. A rise in the price in 1929 caused the reopening of the Conde Auqui mine, thirty miles northeast of Oruro, which proved to be the richest tungsten deposit so far opened in Bolivia. Production in 1929 was 1,630 tons, declined to 410 in 1931, and increased to 1,000 tons in 1932. The actual output is therefore sensitive to the world price, though influenced by exploration and discovery. The Kami mine of the Patiño interests was at one time the largest producer in Bolivia; Chicote, Solano, and Amutara, in the same department, have also produced. In the department of La Paz the Quimsa Cruz, Milluni, Araca, Yaco and Ichoca were at one time or another producers. In Potosí, the Sala Sala, Frias, Llallagua and Uncia should be mentioned, as should the Oruro, Huanani, and San Jorge in Oruro. Most of the tungsten ore in Bolivia is wolframite, but the Conde Auqui and some others produce scheelite. If the world price remains high enough, considerable amounts of tungsten can be produced in Bolivia, which is one of the world's important reserve supply regions.

Antimony ores occur in fissures in Paleozoic shales in the same general regions where the tin deposits are found. Under the influence of the high prices that prevailed during the Great War period, these were worked in a small way by Indians who were financed by ore purchasers, who shipped the product to England. The output, which was only thirty tons in 1913, increased to 4,000 tons in 1914. The industry thus established has persisted and the antimony content of the ore shipped in 1926 reached 4,354.8 metric tons, though it declined to 3,778 metric tons in 1929. This decline seems to have been caused by an increased production of antimony in Mexico, which after 1926 supplanted Bolivia as second world producer. As the world production of antimony for the five years 1926–1930 averaged 28,100 metric tons, of which China produced the major fraction, it is clear that while the output of Bolivia is not negligible it is not likely to increase much unless the Chinese deposits should fail. A trade association in China is endeavoring to control production and price, which will probably be kept at a level where competition from Bolivia does not threaten world markets.

Copper production in Bolivia has long been of some, but not great, importance. About nine-tenths of the production has come from native copper deposits in sandstone which occur in the Corocoro district of La Paz. This is of great geological interest, because it and the Lake Superior deposits of the United States are the only two important regions of the world where native copper is found in workable amounts. A detailed description of their geology has been given by Miller and Singewald and also by L. W. Strauss[4]

One company is the Corocoro United Copper Mines, Ltd., a £750,000 British corporation incorporated in 1909 to take over several smaller companies. This company has a 100-ton concentration plant and a 400-ton flotation plant; in 1919–1920 it produced 14,672,022 pounds of copper. In 1917, it paid a dividend of six francs per share, but has paid none since, and in 1931 the capital of the company was written down to £451,988. A second producing company is the Cia. Corocoro de Bolivia which has been operating since 1873. It has a capital of 1,025,000 bolivianos and is equipped with concentration plant that ordinarily produces about 6,000 tons of 48 per cent copper concentrate yearly. The mine is 1,300 feet deep and employed about 1,000 men when it was working. The high altitude and remote situation of these mines are unfavorable to low working costs and they have been seriously affected by the recent great decline in the price of copper. Small amounts of copper are recovered in connection with the treatment of the silver-tin ores, but are of no great importance. The copper content of the ore, concentrates, *etc.*, exported from Bolivia in 1929 was reported as 7,178 metric tons, while from 1916 to 1923 the production was 10,000 to 11,000 tons yearly.

Lead deposits are widely distributed along the Eastern Cordillera in Bolivia and F. Ahlfeld has published[5] a map on which twenty-nine lead-bearing deposits are shown, ranging from north of Lake Titicaca to near the Argentine border at La Quiaca. They are generally so

[4] L. W. Strauss: *Mining Magazine,* London, September, 1912.

[5] F. Ahlfeld: "Die Bleilagerstätten Boliviens, Freidrich Ahlfeld, Metall und Erz," pp. 265–270, June, 1928.

remote from transportation facilities, of limited size and of such a grade that before the World War but little attention was paid to them. But during the war the price of lead was high and in 1924 and 1925 its price in London was nearly three times what it had been during the ten years preceding 1914. As a result, the exports of lead ore from Bolivia, which in 1913 had been only a little over 1,000 tons but had increased steadily to 9,000 tons in 1923, rose to nearly 15,000 tons in 1925 and to almost 21,000 tons in 1926. This was hand-sorted ore and concentrate averaging 70 to 75 per cent lead. In addition, in 1924, nearly 19,000 metric tons of lead-bearing slag from old silver operations near Potosí and Oruro was exported. These slag exports declined to 16,000 tons in 1925, to 9,725 tons in 1926, and to 2,000 tons in 1927. The lead content of the ore and slag exported from Bolivia in 1926 was 30,900 metric tons, in 1929 it was 21,600 tons, but in 1932 it declined to 5,488 tons.

Nearly all the lead-producing properties are small ones, worked without machinery and with only the simplest equipment, the hand-picked ore and concentrate being transported on the backs of llamas to the nearest railway. Very little margin of profit is left after the heavy transportation charges are paid and it has been estimated that unless lead commands a higher price than five cents gold per pound, very few of them can operate. With improvement of transportation facilities and consequent lowering of costs, some increase of lead production in Bolivia will undoubtely result, but it seems somewhat significant that while some of the larger lead-producing companies of the world have made preliminary investigations of conditions in Bolivia, none of them has, so far as known, yet invested any capital there.

The lead deposits near the Argentine border form an

exception to this generalization, in that they are able to ship their ore to lead smelters in Argentina, where there is, in normal times, a consumption of about 10,000 to 14,000 tons of lead annually and the conditions of supply are such that about 6,000 tons of lead, as metal, or in ore, must be imported. Bolivia is the economic source for lead-ore imports and in recent years practically all the lead ore imported into Argentina was from Bolivia, imports in 1929 being 7,867 and in 1930 6,500 metric tons. This commerce is, therefore, likely to grow.

Zinc is associated with lead and silver in various deposits, the most notable being that of the Compañia Huanchaca de Bolivia, which owns 11,400 acres in the Pulacayo district and is reported to have produced $46,000,000 worth of silver between 1877 and 1892. Recent operations by the Hochschild firm have consisted mainly in treating old dumps with recently minor operations under ground. News reports in 1930 stated that the ore averaged 12 per cent zinc, 4 per cent lead, ½ per cent copper, and sixteen ounces of silver per ton. A 400-ton ore flotation plant was increased in capacity to 700 tons, with about 3,000,000 tons of workable material reported as available. Another zinc producer is the Matilde mine on Lake Titicaca. Relatively little has been published regarding the zinc resources of Bolivia, because they seem to have but little commercial future. Zinc concentrate containing 60 per cent zinc ordinarily sells in the open market in the United States for about 1½ cents per pound, so that it would be only under exceptional circumstances, such as sufficient associated silver or in a period of extraordinarily high prices for metal, that one would expect much production from Bolivia, where transportation costs are necessarily high. Exports in 1926 were reported as 17,560 tons of concentrate, containing 7,601 metric tons of zinc. By 1929, exports had declined to 2,972 tons, but increased to 11,292 tons in 1930, to 26,981 tons in 1931 and had again shrunk to 12,968 metric tons in 1932. If transportation costs could be sufficiently cut, Bolivia has deposits that could considerably augument its present production of lead and zinc. If future petroleum development should give rise to an abundant supply of gas in the eastern plains region, it might possibly be practicable to erect a zinc smelter there, exporting the metal to Argentina, where there is a growing demand for zinc-coated iron articles.

A small amount of coal of Carboniferous age occurs on the Copocabana peninsula, in Lake Titicaca. According to a published description of the deposit, it is not likely that any bed exceeds 300 or 400 feet in length and three feet in thickness. About 1875, the deposit was worked, producing about thirty tons daily, but the mine has long been idle. It is unfortunate that the geology of Bolivia is such that it is not likely that workable coal deposits will be found there, for wood fuel is exceedingly scarce on the high plateau. The natives simply bear the low temperatures, but foreigners are exceedingly uncomfortable in the unheated dwellings. The absence of fuel makes any metallurgical operations expensive and impracticable. It will have been noticed that all the minerals are exported as ore or concentrate for reduction abroad.

The Salta-Jujuy petroleum field of Argentina extends across the border into Bolivia and indications are that petroleum occurs as much as 150 miles north of the border. The Standard Oil Co. of Bolivia, a subsidiary of the Standard Oil Co. of New Jersey, entered into an agreement with the Bolivian government under which it was obligated to begin production by 1930. At the end of that time it had completed five wells in the Sanandita field, of which the first two were producers, the third and fourth poor and the fifth produced nothing. In the Bermejo field five wells were drilled, but no statement regarding them has been given out. Three wells were also drilled farther north, at Camiri, Camatindi, and Maracheti. The last produced nothing, and the results on the others are not known. The original Levering concession, which was taken over by the Standard Oil, provided for a thirty-mile right of way for a pipe-line across the Chaco, but as that region has been under dispute between Paraguay and Bolivia for a hundred years and the locus of a war since 1931, none has been built.

The only other American company interested in Bolivia is the Bolivian Petroleum Corporation, which originally had a concession to 1,250,000 hectares on the Altiplano. Of this it retained only 125,000 hectares, and acquired 250,000 hectares to the east along the route of the railroad being constructed from Cochabamba to Santa Cruz. Apparently it has not yet drilled any wells. A local company was formed in 1930 to exploit 1,500,000 hectares near Cochabamba.

During 1930, a new petroleum law was passed by Bolivia in which the country is divided into four zones, with taxes during the period of geological exploration ranging from one centavo in the fourth zone to two and one-half centavos in the first zone. A registration fee ranging from one-quarter centavo per hectare in the fourth zone to one centavo in the first zone was substituted for the deposit of guarantee previously required, which had proved unsatisfactory. Provision is made for the number of wells which must be drilled yearly in the period of exploration and the period of production. Taxes in the latter period range from three centavos to six centavos per hectare during preparatory drilling. When the company begins production, its taxes rise on a sliding scale for seven years, reaching fifty centavos per hectare in the seventh year for the first zone. State royalties are 12 per cent in zone 1, 11 per cent in the second zone, and 10 per cent in the fourth zone. Oddly enough, the royalty decreases with increasing production, being only 7 per cent on an output of 10 million barrels per year. The petroleum industry is classed as a public utility and may be expropriated by the government. The new law, which is much more satisfactory than the old one, has not yet gone into effect.

Imports of petroleum and its products into Bolivia have ordinarily amounted to about $1,300,000 yearly, but the requirements will increase as automobile roads are extended. The total highway mileage in 1930 was 3,875, with 992 miles additional under construction. One of the most marked results of the activities of the Standard Oil of Bolivia was the construction of a network of motor roads by which journeys which formerly required ten to fifteen days on muleback can be made in two or three days by motor car in the dry season. In recent years, the price of gasoline (from Peru) has been fifty to sixty cents per gallon in Bolivia and the price of fuel oil about $8 per barrel.

It is possible that Bolivia may eventually make a substantial contribution to the petroleum supply of the world, but any extensive development must await a permanent settlement of the political questions now agitating the governments of Paraguay and Bolivia, and the establishment of a stable government in the latter country.

Production of quicksilver in Bolivia was started in 1928, exports of 1,575 kilograms in that year and of

1,741 kilograms in 1929 being reported. Nothing has been published regarding this enterprise; possibly it represents cleaning up old Spanish silver residues.

The only other mineral export of Bolivia reported in recent years was 24,665 kilograms mica in 1926, 16,717 kilograms in 1927, and 11,285 kilograms in 1928. In 1929, the exports declined to 1,986 kilograms. In 1927, the export of 209 kilograms of asbestos was reported, but none in 1928 or 1929; most of the occurrences are not of commercial quality.

No information seems to be available as to other minerals in Bolivia. One would expect to find sulphur deposits and other non-metallic minerals, but nothing seems to have been published concerning them. With transportation and living conditions as unfavorable as they are, there seems little likelihood that mineral deposits of low unit value can possibly be produced for export.

That leads into the difficult subject of the character and probable future development of the population of Bolivia. It has already been stated that the population is over 50 per cent Indian and 30 per cent mestizo. Less than 15 per cent of the population controls the best lands, the wealth, the education and the politics of the country. The managers of mines speak well of the Indian, praising his docility, honesty, strength and animal-like capacity for continuous hard labor under difficult conditions. Others refer to his excessive use of alcohol and coca, which seem to produce a condition almost of stupor. He is subject to disease and lives in unheated huts upon poorly prepared food of the simplest character. There is even more disagreement regarding the cholos, as persons of mixed blood are called. The published statement by a visiting engineer that "The cholo seems to have retained many of the worst characteristics of the natives and the

Spaniards from whom he sprang, and to my mind is less likeable than the native, although it is treason to say so" gave rise to much resentment but no very effective rebuttal. People of mixed blood, of course, show a wide range of characteristics, and it is not our purpose here to attempt an accurate social evaluation of the Bolivian, but merely to make the point that the general economy of Bolivia is such that only a small fraction of its population live in a way which makes any appreciable demand on mineral commodities. The natives made but little use of minerals before the Spaniards arrived and would still make but little use of them were the mineral-producing industry taken out of Bolivia. Extensive as that industry is, nothing is more certain than that it will eventually decline. San Francisco and Sacramento, originally towns based on a mining boom, have struck their roots into a normal economy and flourished; Potosí, like Virginia City, has dwindled with the decline of nearby mining. There will undoubtedly be additional mineral discoveries in the high plateau of Bolivia, and the tin-producing industry, under cartel regulation of output, will long endure, but on the whole it would seem that the highlands will never be much different from their present state.

The undeveloped eastern lowland is quite a different matter. Held back from development by lack of transportation and political uncertainty, mineral enterprise has already made a start toward improvement of transportation facilities, and will probably go much farther if the political difficulties can be settled. The Sacramento of the future in Bolivia, if it ever develops, will be in the eastern lowland instead of upon the high plateau. In the immediate future Bolivia will remain what it has been for four centuries, a country whose economy is based on the production for export of minerals of high unit value.

CHAPTER XII

ECUADOR

ECUADOR is perched on the backbone of the Andes like a saddle on a high horse. It has an area variously estimated as 110,000 to 275,000 square miles, the most probable figure being 167,000; the Galapagos islands also belong to it. With a frontage of about 600 miles on the Pacific, the country extends to the east in the form of a triangle, with the apex of the disputed area on the Amazon, 680 miles to the east. The country is wholly in the tropics, extending from less than 2° N. latitude to little more than 4½° S. Fortunately, however, nearly half of the area is occupied by the mountains and high plateau of the Cordilleras, so that it enjoys a temperate climate. In Bolivia and Peru, to the south, the Andes widen out into a great tableland with a number of separate ranges. In Colombia, to the north, they again divide into three main and several subordinate ranges, but in Ecuador the whole Andean system is compressed into what is essentially one range, though with an eastern and a western crest and a series of basins between which collectively form a long central upland. The mountain mass as a whole is some 500 miles long and averages perhaps 100 miles in width. The central upland is some 7,000 to 10,000 feet high, some with passes above 11,000 feet through the high mountain crests to the east and west.

Many volcanoes rise typically 3,000 to 4,000 feet above the plateau, reaching a climax in Chimborazo at 20,576 feet. Approximately three-fourths of the population live on this mountain upland and in places there is a population density of thirty to sixty per square mile, as high as that in the most favored portions of South America. The economy of the country is essentially self-contained, the external trade per capita amounting to only one-seventh that of Uruguay, the smallest country on the continent. The great elevation makes the climate of this populous portion of the country attractive and an unusual variety of agricultural products are grown. The people are engaged in the production of those goods, mostly food and clothing, that a simple agricultural type of economy requires. No accurate census has ever been taken and the population is estimated variously at 1,500,000 to 2,000,000. It consists of whites, Indians, and mestizos, the Indians constituting the great bulk, and the proportion of whites being estimated as low as 1 per cent. The country exports $15,000,000 to $20,000,000 worth of goods per year and imports $11,000,000 to $17,000,000 worth. Invisible items change this favorable balance in goods to an unfavorable one. Of the exports, a quarter to a third are ordinarily derived from cacao beans, an eighth to a fifth from petroleum, a tenth to a fifth from coffee, and 8 to 10 per cent each from "Panama" hats, and gold and silver; but in 1931 petroleum corresponded to over a quarter of the total exports, nearly equalling cacao and coffee combined. The public debt at the end of 1930 was $24,866,520, about $10 per capita.

Physiographically, the country may be divided into three regions: the Oriente, the Andes, and the coastal plain. From the base of the eastern range, at about 4,000 feet altitude, the country slopes down to the lowlands of the Amazon basin at 850 feet. This part amounts in area to nearly half of the total. Connected with the upland by only a few perilous trails, the Oriente is inhabited by less than 100,000 people, most of them pure-blooded natives, who neither wish nor have contact with white men. A few small towns lie at the base of the mountain range, and a little rubber is collected for shipment down the Amazon under conditions paralleling those of the fur trade of our own northwest in 1800. Its eastern portion merges into the 100,000-square-mile area which has been for many years a cause of jurisdictional dispute among Brazil, Colombia, Peru, and Ecuador. About half of Ecuador is really a sort of vacant lot for which there seems little prospect of any considerable development, either in the near or more remote future, though as noted elsewhere it contains possible petroleum territory.

West of the Andes there is a much smaller area of perhaps 30,000 square miles. Though described in geographies as the place at which the Pacific coastal plain reaches a width of 80 to 100 miles, it is hardly a plain save by contrast with the Andes, being occasionally intersected by mountains that rise as high as 3,000 feet and rising along its eastern edge at the foot of the Andes to a general elevation of 1,000 to 1,700 feet. To the south, parts of this area have a climate like that of the deserts of Peru, but most of it is warm and rainy. Here grow the cacao trees, whose beans are the raw material for chocolate. Long the world's principal source of this important food product, a disease which attacks the pods broke out in 1916 and was followed in 1922 by another, known as "witch broom," which damages the trees. Exports in 1928 dropped to the level of 1890, a serious thing for Ecuador, since in many years cacao represented three-quarters of its total exports and had been a source of revenue to the country much as nitrate had been for Chile. Disease-resistant varieties of trees will presumably be developed eventually but meanwhile there is danger that cacao-growing elsewhere in the world will be so stimulated that Ecuador will never regain its dominant position.

Except for gold and petroleum, minerals have never been of much importance in Ecuador. The Spanish invaders found the natives of the southern part of the country washing placer gold, and in 1549 established a town at Zaruma, which is in the central plateau, near its southern end. At another place to the southeast and at two more farther north mines were started about this time, but an uprising of the natives, about 1600, brought organized operations to an end. They have not been resumed except at Zaruma. Silver mines apparently were worked in the early days but no attempts to reopen them have been made for the past forty years, and the 60,000 to 90,000 ounces of silver output reported yearly is practically all incidental to that of gold, with which it is associated in the Zaruma mine. Some copper is associated with silver ores, but there is no production, nor any of lead; zinc has never even been reported. Quicksilver minerals are known to occur but do not seem likely to become the basis of commercial production. It is claimed that one mine was worked by the Spaniards, but, if true, they removed the ore so completely that it is impossible now to determine what it was. The production of some sulphur is reported from beds in the marine Pleistocene deposits in the Santa Elena oil fields of the coast. Sulphur also occurs around the volcanoes and in the Galapagos islands, but the production is unimportant. Rock salt is reported to occur in the oil fields, but all the recorded production is obtained by solar evaporation of seawater or from brines from springs.

Petroleum in Ecuador is of unusual interest because the oil springs of the Santa Elena peninsula west of Guayaquil seem with little doubt the locus of the first discovery of petroleum in the new world. A few hand-dug pits were worked by the natives when the Spaniards entered the region. These yielded enough for the simple needs of the time and the Spaniards did not add to the requirements. Though dating back so early, no important development resulted and as late as 1917–1922, annual output ranged between 50,000 and 60,000 barrels yearly, all derived from a thousand or more shallow pits. More systematic attempts at development then began to show effect; the output in 1924 reached 100,000 barrels and doubled each year thereafter, reaching 1,084,000 in 1928. In 1929, it showed only a 35 per cent increase, reaching 1,350,000 barrels, and in 1930 was 1,559,000 barrels. Of the amount produced in 1926, a single company, the Anglo-Ecuadorian (which, as its name indicates, is English) accounted for almost 90 per cent from its fifteen or twenty drilled wells. Of this company's production, about 100,000 barrels were refined and marketed in Ecuador, and the rest exported. The South American Gulf Oil Corporation, International Petroleum Co., Ltd., and the Standard Oil of California have also drilled wells in this region. Exports in 1931 were 1,381,121 barrels, valued at $3,165,000.

In 1921, the Leonard Oil Development Co. was granted a concession covering about 10,000 square miles in the region east of the Andes. It will readily be understood that actual exploration for oil in this region can be carried on only with extreme difficulty and results may be slow in materializing, but the outlook for possible

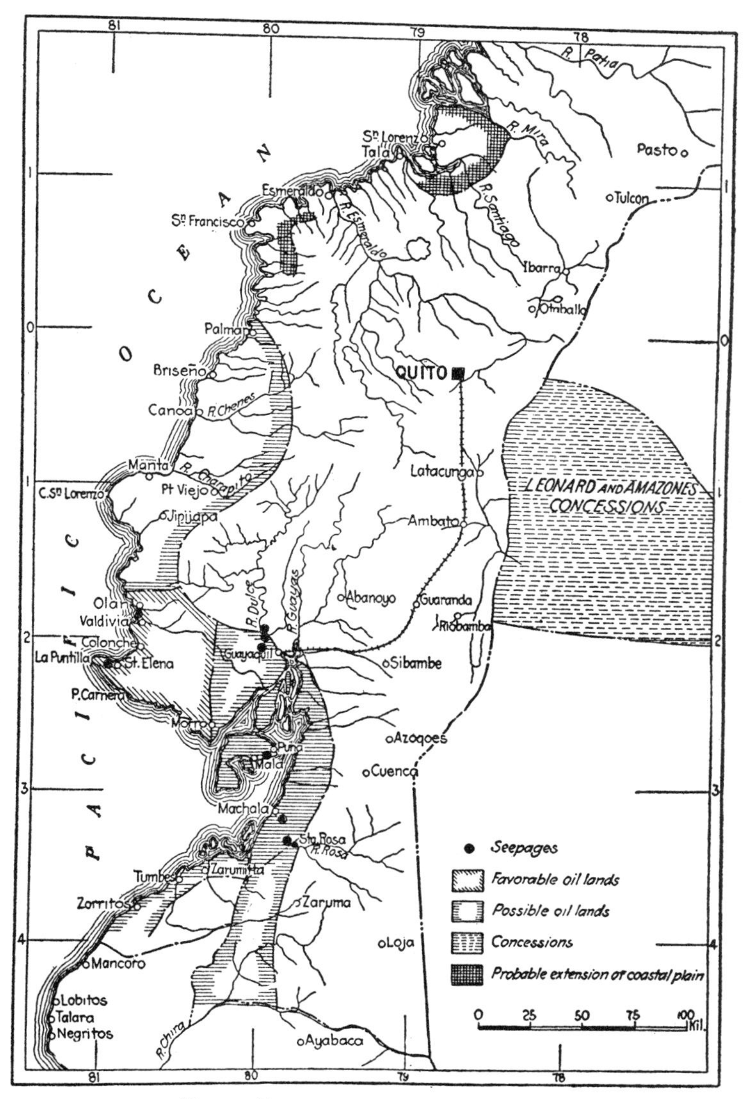

FIG. 7. PETROLEUM REGIONS, ECUADOR

large production there is more favorable than in the already producing region. If and when it comes about, it will be accompanied by improved transportation facilities that will open the way for general economic development of the region. As a factor in world production of petroleum, Ecuador is not likely to be of much importance, for many years at least, but as a factor in the internal economy of the country oil seems certain to rank first of mineral substances.

Coal mining is unimportant, though in proportion to its size more coal resources are claimed for Ecuador than for most South American countries. Both on the west slope of the Andes and in the intermountain valley, rocks of Tertiary age occur and beds of lignite are found in the series. The Tertiary rocks on the west slope are mostly marine sediments and such bituminous deposits as occur are usually not workable. Those in the intermountain region are of better quality, but even there they are high in ash. A lignite deposit near the railway to Quito from Guayaquil was once expected to be a source of coal supply for the road, but proved on exploration to have been tilted into a vertical position and to be practically unworkable. Deposits at various other places in the intermountain region would involve the construction of branch lines. The limited market for coal and a domestic supply of crude oil, of which there is a surplus for export, make the investment of capital in the production of a low-grade solid fuel like lignite commercially unattractive. During the decade preceding 1919, Ecuador imported about 30,000 tons of coal annually. Ecuadorean coal cannot be produced for export, or even to compete with imported fuel in the ports, so that whatever development eventually takes place can hardly exceed a small production for the limited markets that exist in the intermountain region. It has already been remarked that except for petroleum, cacao, and a few other commodities, production in Ecuador is mostly to meet domestic needs. Though possessing the fuel, both liquid and solid, to permit the development of mechanical production, economic forces do not seem to be trending in that direction.

There seem to be no reports of the existence of iron-ore deposits, and what has been said above explains why there is little economic pressure toward the development of a domestic iron industry. Petroleum production involves considerable requirements of iron pipe and machinery and these will undoubtedly continue to be supplied through imports. Ecuador has less than 600 miles of railway; here again the metal needs can best be met by imports.

Gold is obtained from both placers and lode deposits. The placers of the Santiago river system, in the northwest corner of Ecuador, have been yielding a small production ever since the sixteenth century. An estimate[1] made in 1906 credited this region with an annual output of between $200,000 and $300,000. After 1890, much interest was aroused in this region, various English and American companies acquired lands, and production on a larger scale was developed. None of the ventures proved profitable and by 1919 only small-scale native work was still in progress. The total gold production in Ecuador in 1928 was reported as 74,566 ounces and as the lode mine next described accounted for 70,931 ounces, the total for the rest of the country was evidently not large. The occurrence is of interest in that general conditions are similar to those just to the north in Colombia and some platinum also occurs with the gold.

[1] L. L. Welmire: "Gold Dredging in Ecuador," *Mining Magazine*, vol. 6, pp. 385–391, 1906.

The Zaruma gold mine is in the intermountain region, near the southern border, forty-five miles inland by trail from the little port of Santa Rosa and in a region where the elevation varies from 2,000 to 5,000 feet. Part of the distance will be traversed by a railroad now in course of construction. The Spaniards began working there, following the lead of the natives, in 1549, and some production at least was maintained until it was interrupted by the war of independence. In 1880, the Great Zaruma Gold Mining Co. was organized in London but went bankrupt attempting to construct a road and move machinery in from the coast. The Zaruma Gold Mining Co. was organized in 1887. In 1892, an American company, the South American Development Company, took over the properties and has worked them steadily ever since. Detailed descriptions of the property have been published[2] and it does not seem necessary to say more here than that faulted but persistent veins of fair size yield ore that carries about one-half ounce gold and a slightly larger amount of silver per ton. The enterprise furnishes direct employment for about 2,000 citizens of Ecuador and indirectly for a much larger number engaged in supplying the various needs of the group directly employed. It is also important as furnishing a small but steady domestic output of gold.

The influence that minerals have exerted in the development of Ecuador can easily be summarized. In Colonial days, the search was, as usual, for gold and silver, with incidental attention to lead. The considerable quantity of gold articles in the possession of the natives created a natural hope of finding rich and large deposits of gold.

[2] J. W. Mercer: "Mining in Ecuador," Second Pan American Scientific Congress, Washington, 1916. Reprinted, *Eng. and Min. Jour.*, New York, February 19, 1916.

These hopes proved baseless, for, although gold was found in a number of places, the deposits were neither large nor rich. Mining operations, which had never been extensive, were, as already noted, brought to a practical stand-still by the war of independence, and did not greatly revive until about fifty years ago, when there was a recrudescence of interest in gold and silver mining. The net result of this has been the development of one fair-sized persistent gold mine, although considerable amounts of foreign capital have been expended elsewhere in the country on fruitless attempts at development. Petroleum production, which had existed on a small scale from the beginning, remained small until the period of the Great War. Between 1923 and 1928, annual production increased ten times, and while the total is not large it is important for so small a country as Ecuador, since it exceeds domestic requirements by more than ten times and furnishes an important export commodity. The present development is probably near commercial equilibrium and no further marked increases can be expected unless large fields should be found in the unexplored region east of the Andes. If that should occur, it will lead to the general economic development of a large fraction of the total area of Ecuador which has as yet remained wholly undeveloped, much as the discovery of gold in California led to rapid development there.

So far, mineral development has had little effect on the industrialization of the country, though the mines and the oil fields have trained mechanics and mineral exports have created foreign credits. Perhaps $25,000,000 of foreign capital has been invested in mineral production and some of the shorter railway lines have been built in expectation of mineral freight. In all, some 10,000 people have been employed in the mines and oil fields and they, and others through them, have been brought into contact with new standards of living. No notable controversies over mineral rights have disturbed the foreign relations of the country, and special laws governing mineral development are not oppressive. There is a reasonably open field for the development of such resources as are present.

Ecuador possesses coal resources, but the quality is poor, mining conditions are difficult, and the scattered and small markets are not favorable to the development of coal-mining enterprise, especially in the face of an abundant supply of domestic crude petroleum. Except for small local development, the coal industry is not likely to grow. Other forms of mineral enterprise are but little developed, for the general economy of the country as a whole is a rather simple agricultural one. Except for the oil fields of the Santa Elena peninsula, the modern gold mine at Zaruma and the railroads serving the towns and small cities (the largest does not exceed 50,000 population) of the intermountain region, the country remains essentially an agricultural region. The opening of the Panama canal, and more especially the eradication of yellow fever in the port of Guayaquil after 1919, have greatly aided social progress in Ecuador, and steady development may be expected, but with the sole exception of petroleum it is not likely to become an important mineral-producing country.

CHAPTER XIII

WHO OWNS THE MINES OF SOUTH AMERICA?

IF ONE judges by the present volume of production, most of the mineral resources of South America are owned by foreign capital, American investment in particular being dominant. There are two notable exceptions: Bolivia, where the main ownership of the tin industry is domestic, and Chile, where the nitrate industry was until recently controlled by the government. Even in these instances there is a substantial participation by foreign investors. In making this study of ownership, many small properties have unavoidably been neglected.[1]

Antimony.—Within recent years, Bolivia has become the second country in the production of antimony. The two principal sources of production have been those owned by Tramonti and by H. Gerike, while the properties of Ricardo Cruz and J. M. de Recacochea have supplied lesser amounts.

Asphalt.—Although most commercial asphalt is now a by-product from the refining of petroleum, an American company owns the two principal sources of natural asphalt, one of which is in Trinidad and the other in Venezuela. The General Asphalt Co operates in Trinidad through its subsidiary, the Trinidad Lake Asphalt Oper-

[1] This chapter has been prepared as a summary by W. P. Rawles, Secretary of the Mineral Inquiry, of data in his general bulletin, "The Nationality of Commercial Control of World Minerals," New York, 1933.

ating Co., Ltd., and in Venezuela through another subsidiary, the New York and Bermudez Co. There is a smaller production under British control and at one time a German company undertook development in Venezuela.

Bauxite.—British Guiana has produced about 10 per cent of the world's annual output of bauxite during the last few years. The Demerara Bauxite Co., Ltd., a subsidiary of Aluminium, Ltd., a Canadian corporation whose stock is largely held in the United States, has produced practically all of the tonnage. The British and Colonial Bauxite Co., Ltd., a British concern, secured leases on bauxite properties in 1929. In Surinam, the production of bauxite has been about the same as that in the neighboring British colony. There are two companies operating, the Kalbfleisch Corporation, a subsidiary of the American Cyanamid Co., and the Surinaamsche Bauxite Mattschappij, a subsidiary of the Aluminum Co. of America. The latter company has accounted for about nine-tenths of the production in recent years.

Bismuth.—A Swiss-Bolivian company, Cie Aramayo des Mines en Bolivie, has been the chief producer in Bolivia. In Peru, the San Gregorio mine of the Negociacion Minera Fernandini, a domestic company, was long the principal source of bismuth, but the Cerro de Pasco Copper Corp., an American company, has now become the dominant interest.

Chromite.—The mine at Santa Luzia has been the chief source of chromite in Brazil and in 1927, the last year of production, it was operated by E. J. Lavino and Co., an American concern.

Coal.—Very little coal is mined in Brazil and that from properties owned by local capital, although an American utilities company, American and Foreign Power, is interested in one mine. Chile has the largest production of coal of any of the South American countries, but, at that, it is a small fraction of 1 per cent of the world's total. The coal mines are owned by domestic capital with the exception of the small Mafil mine which is owned by an American utilities company through the Soc. Carbonifera de Mafil. In Colombia, the State Railways produce most of the coal mined but there are small mines serving local industries. The ownership is domestic. In Peru, the Cerro de Pasco Copper Corp. produces four-fifths of the coal, and another American company, the Northern Peru Mining and Smelting Co., a subsidiary of the American Smelting and Refining Co., furnishes most of the remainder. In Venezuela, the coal mines are owned by domestic corporations.

Copper.—The production of copper in South America is negligible outside of Chile and Peru. In Argentina, in spite of the effort to revive the Corporacion Minera Famatina with national funds, there is no output. In Bolivia, a company controlled by French capital, Corocoro United Copper Mines, has accounted for most of the copper production in recent years, but ceased mining in 1930.

Chile is the world's second copper-producing country and supplied over 16 per cent of the total production in 1929. The Chile Copper Co. and the Andes Copper Mining Co., both subsidiaries of the Anaconda Copper Co., produce about two-thirds of the copper in Chile. The Braden Copper Mines Co., a subsidiary of the Kennecott Copper Corp., supplied nearly one-fourth. The remainder was produced by the Poderosa Mining Co., Ltd., a British company, Soc. des Mines de Cuivre de Naltagua, a French company, and various Chilean companies, including Cia Minera de Tocopilla, Cia Minera Disputada de los Condes, and Cia de Mines

Gatico. The properties of the Chile Copper Co. contain more than 27 per cent of the world's known reserves of the metal, those of the Braden Copper Mines Co. 5 per cent, and those of the Andes company 2 per cent. About 3 per cent of the world's copper comes from Peru, the principal producer being the Cerro de Pasco Copper Corp., already mentioned, which accounted for four-fifths of the output in 1929. The Northern Peru Mining and Smelting Co., a subsidiary of the American Smelting and Refining Co., furnishes most of the remainder.

In Venezuela, the South American Copper Mines Syndicate, a British company, during its intermittent periods of operation, has produced practically all of the copper shipped.

Gold.—In Argentina, the output is small and irregular. It comes from mines under Argentine control. In Bolivia, while the gold production has been unimportant in the last few years, confident predictions have been made for a good yield by an American company, the Bolivian Gold Exploration Co. In Brazil, practically the entire output of gold comes from the famous St. John del Rey mine, owned by a British mining company of that name. In British Guiana, only one gold-dredging company operates, the Minnehaha Development Co., Ltd., a British concern. This company produced one-half of the gold in the colony, the remainder being washed from alluvium by native laborers. The bulk of the Chilean gold comes from the base metal mines and, in 1929, more than one-half came from the operations of the Andes Copper Mining Co. Cia Minera Las Vacas was reported as being the only gold mine in operation in 1931.

In Colombia in recent years, the Pato Mines, Ltd., has produced nearly one-half the annual output of gold, and another British firm, Frontino and Bolivia Gold Mining

Co., about one-third. The Cia Minera Chocó del Pacifico, S. A., the local operating company of the South American Gold and Platinum Co., an American concern in which there is an important British interest, has accounted for one-sixth, and a native company, Soc. de Zancudo, has supplied a small amount. The South American Development Co., an American company, produces practically all of the gold in Ecuador. Virtually all of that from French Guiana is produced by a French company, Soc. Nouvelle de Saint-Elie et Adieu-Vat. There are no companies actively engaged in gold mining in Surinam, the entire production being obtained from the washing of alluvium by native laborers. The French company, Cie. des Mines d'Or de la Guyane Hollandaise, has been taken over by a British company, Neotropical Concessions, Ltd.

Peru is the largest producer of gold in South America although, in 1929, the country supplied only about one-half of 1 per cent of the world's total production. The Northern Peru Mining and Smelting Co. produced nearly one-half of the output in 1929 and the Cerro de Pasco Copper Corporation about one-fifth. The Inca Mining and Development Co., also American, supplied about one-tenth. Another tenth was produced by La Cotatambas Auraria Mina, a Peruvian company, The Anglo-French Ticapampa Silver Mining Co., whose nationality is indicated by its name, produced an appreciable quantity. The efforts of individual laborers also yielded a substantial amount.

The principal gold-mining companies in Venezuela are British. The Botanamo Mining Corporation accounted for nearly one-half of the output in 1930 and the New Gold Fields of Venezuela, one-quarter. The Burdette Mining and Development Co., an American company, and The Cia Francesca la Macupia, a French concern, also operate in the country near New Gold Fields of Venezuela.

Iron.—Although the present production of iron ore in Brazil is negligible, the reserves are famous for their size and quality. The principal holdings are divided between the Itabira Iron Ore Co., Ltd., a British firm, and the Brazilian Iron and Steel Co., an American concern. Both have holdings in the Itabira district and the American company has additional holdings in the São Miguel and Santa Rita districts. Another American company also has holdings in the Santa Rita district. A Belgian syndicate, Conpanhia Siderurgica Belga-Mineira, at present the principal producer of pig iron in Brazil, has small holdings of ore reserves.

Practically all of the iron ore produced in Chile comes from El Tofo, where it is mined by the Bethlehem-Chile Iron Mines Co., a subsidiary of the Bethlehem Steel Co. Among the undeveloped properties in Chile, the best known is Algarroba, owned by a Dutch-German syndicate, controlled by Wm. H. Müller and Co. and Gutehoffnungshütte. The largest known iron-ore deposit in Venezuela is owned by the Bethlehem corporation.

Lead.—The Cia Minera y Metalurgica Sud Americana, a subsidiary of the National Lead Co. of Argentina, which in turn is owned by the National Lead Co., an American company, mines practically all of the lead in Argentina. This company smelts its own ore and controls another company, La Cia Hispano Argentina de Minas y Metales, that treats lead ore imported from Bolivia. Lead will be produced in the future from a zinc-lead mine, owned by Cia Minera Aguilar, S.A., a subsidiary of the St. Joseph Lead Co., an American firm.

The lead production of Bolivia appears to be controlled by Chilean-French capital, a large portion being the output of Cia Huanchaca de Bolivia and the remainder being scattered among small producers. The Matilde mine, owned by M. Hochschild, is a potential producer of some importance.

The major part of the lead mined in Peru in 1929 came from the mines of Negociacion Minera Fernandini, a Peruvian company, and practically all the rest from the properties of Soc. Minera Backus y Johnston del Peru, a subsidiary of the Cerro de Pasco Copper Corporation. Nearly all of the smelter production, during the same year, came from the Cerro de Pasco company's plants and a small amount from the Vesubio and Pompei establishments, both of which are locally owned.

Manganese.—Brazil ranks fourth as a producer of manganese with about 9 per cent of the world's production, according to the figures for 1929. More than one-half of the national output was produced by a subsidiary of the United States Steel Corporation, the Cia Meridional de Mineracão. Other producers of importance are the Santa Mathilde and A. Thun & Co., Ltd. Very little manganese is produced in Chile and most of that has come from the W. R. Grace Co., an American company, the remainder being divided between the firms of Hochschild and Varela.

Mercury.—A vigorous but unsuccessful effort was made to develop the properties in Peru of Negociacion Minera Fernandini as a source of mercury.

Mica.—Mica is produced in small quantities in Argentina and Brazil from properties that are locally owned.

Nitrate.—Almost the entire Chilean nitrate industry was consolidated in Cia de Salitre de Chile (Cosach) recently dissolved. One-half of the stock (all of Class A) in the company was owned by the Chilean government and the other one-half (Class B) was exchanged for stock in the former operating companies. Two-thirds of the Class B stock went to companies that had been controlled by the Guggenheim Brothers in which American and British capital are dominant.

Petroleum.—The largest producer of petroleum in Argentina, the Yacimientos Petroles Fiscales (Y.P.F.), a company operated by the government, supplied about 40 per cent of the national output. Most of the remainder is fairly evenly divided among four groups of companies; the "Diadema" group, controlled by the Royal Dutch Shell combine, "Astra" and "Orienta" Cia Argentina de Petroleo, both under the same German ownership, Cia Ferro-Carrilera de Petroleo, a company operated by the British-owned railways in Argentina to supply fuel oil for their own use, the Standard Oil group including the Standard Oil Co. of Argentina, Cia de Petroleo La Republica, Ltd., and Cia Nacional de Petroleo, all subsidiaries of the Standard Oil Co. of New Jersey. There are three small companies whose production is less than 5 per cent of the country's output. They are S. Petroleo de Challaco, Ltd., a subsidiary of Mines et Petroles Sud-Americains, a Belgian company; the Solano Co., a British concern; and Cia Industrial y Commercial de Petroleo, a subsidiary of the Anglo-Persian Oil Co. in which the British government has a large interest. The largest refinery is that of Y.P.F., those of the Standard, the railways and the Dutch Shell companies being next in importance.

A very little production has been obtained from the large concessions owned in Bolivia by the Standard Oil Co. of Bolivia, a subsidiary of the Standard Oil Co. of New Jersey. Another American company, Bolivian Petroleum Co., owns a large concession but has no producer wells.

The entire output of Colombia in 1930 came from the operations of one company, the Tropical Oil Co., owned by the International Petroleum Co., Ltd., which is controlled by Imperial Oil, Ltd., and it in turn by the Standard Oil Co. of New Jersey. The refining operations of the country are carried on by the same company. Two other American companies have been carrying on development work in the country, the Richmond Petroleum Co. of Colombia, a subsidiary of the Standard Oil Co. of California, and the South American Gulf Oil Corp., a subsidiary of the Gulf Oil Corp. of Pennsylvania. The Gulf company holds the famous Barco concessions from which rich returns are anticipated. Additional lands are held by the Texas, Sinclair and other companies or syndicates, some financed by local capital and others by foreign.

The small production of petroleum in Ecuador during 1930 was dominated by British companies. About 97 per cent of the output came from the wells of the Anglo-Ecuadorian Oilfields, Ltd., another 2 per cent from the properties of the Ecuador Oilfields, Ltd., and a few barrels from the Concepcion Ecuadorian Oilfields. The International Petroleum Co., Ltd., already mentioned, holds large concessions but did not obtain an appreciable production in 1930. There was a slight production by native companies. The Anglo-Ecuadorian Oilfields, Ltd., has the only refinery in the country.

Over three-fourths of the petroleum produced in Peru in 1930 came from the wells of the International Petroleum Co., Ltd. A British company, La Cia Petrolera Lobitos, supplied practically all of the remainder. The

International Petroleum Co., Ltd., has a refinery capable of treating almost one-half of the national output of crude oil. Another company, Establecimiento Industrial de Petroleo de Zorritos, which is owned locally, has a small capacity.

British companies produce more than 80 per cent of the crude oil in Trinidad. The two that are the most important are Apex Oilfields, with more than 30 per cent of the production of the colony, and Trinidad Leaseholds, Ltd., with 13 per cent. The other British companies operating are the Kern Trinidad Oilfields, Venezuelan Consolidated Oilfields, Trinidad Petroleum Development Co., Ltd., British Controlled Oilfields, and Petroleum Options, Ltd. United British Oilfields, a subsidiary of the Royal Dutch Shell combine (40 per cent owned by British capital), produces about the same amount as Trinidad Leaseholds, Ltd. Two American companies, Trinidad Lake Petroleum Co. and Petroleum Development Co., Ltd., both subsidiaries of General Asphalt Co., produce about 4 per cent of the petroleum on the island and a third American unit, Trinidad Oilfields Operating Co., Ltd., a subsidiary of the Standard Oil Co. of New Jersey, about 2 per cent. There are four refineries on the island with a combined capacity just about equal to the petroleum production. That of Trinidad Leaseholds, Ltd., is the largest, with more than 80 per cent of the total capacity, while the plant of the United British Refineries, a subsidiary of the Dutch Shell combine, has about 15 per cent. The West India Oil Refining Co., a subsidiary of the Standard Oil Co. of New Jersey, and a British company, Trinidad Central Oilfields, have small plants.

Venezuela is the leading petroleum-producing country in South America and for a while ranked second in the world. While the official reports of the Venezuelan government list many companies as operating in the country, there are only about one dozen that have production of any consequence and only four that are large. Three companies, all subsidiaries of the Royal Dutch Shell group, produced nearly one-half of the oil in Venezuela in 1930. Venezuelan Oil Concessions, Ltd., is the largest company operating in the country, accounting for 29 per cent in that year, the Caribbean Petroleum Co. about 15 per cent and the Colon Development Co., Ltd., less than 4 per cent. The second most important group is that controlled by the Standard Oil Co. of New Jersey. In 1930, these companies produced 29 per cent of the crude oil in Venezuela. Nearly all of this was produced by the Lago Petroleum Corp., which is controlled by the parent company through the Pan-American Petroleum and Transport Co. The others are the Standard Oil Co. of Venezuela, with its subsidiary the American British Oil Co., and the Rio Palmar Oilfields, all controlled through the Creole Petroleum Corp. In the following year, the Standard Oil Co. of Venezuela increased its production considerably. The Venezuela Gulf Oil Co., a subsidiary of the Gulf Oil Corp. of Pennsylvania, produced about 21 per cent of the petroleum in Venezuela in 1930. The remaining companies furnish a very small fraction of the total production. There are three British companies, British Controlled Oilfields, Tocuyo Oilfields of Venezuela and the Central Area Exploitation Co. of Venezuela. The remaining American companies are the Bermudez Co., a subsidiary of General Asphalt Co., the Coro Petroleum Corp., the Richmond Petroleum Corp. of Venezuela, the former an affiliate and the latter a subsidiary of the Standard Oil Co. of California.

While there is no petroleum production in the Dutch

West Indies, a large amount of the crude oil from Venezuela is refined here. A subsidiary of the Royal Dutch Shell operates a large refinery on the island of Curaçao and a small one on Aruba, where the Lago company also has a large plant.

Platinum.—A third or more of the world's platinum is produced in Colombia, the only countries with a larger output being Russia and Canada. Through its operating company, Cia Minera Chocó Pacifico, the South American Gold and Platinum Co., an American concern, produces slightly more than one-half of the platinum in the country. A Colombia company, Intendencia del Chocó, produces practically all of the rest, although there is a smaller native company, Costa del Sur, and a small British company, the British Platinum and Gold Corp.

Silver.—Silver is present in the lead ores mined in Argentina by Cia Minera y Metalurgica Sud America, which is controlled by an American company, the National Lead Co. During its intermittent periods of operation, the Famatina company has produced silver and the Cia Minera Aguilar, a subsidiary of the St. Joseph Lead Co., is expected to obtain silver from its zinc-lead mine. Practically all the silver from Bolivia is produced by two companies. Cie Aramayo des Mines en Bolivie, a company incorporated in Switzerland but controlled by Bolivian capital with considerable British and American participation, produced about three-fifths of the national output. Cia Huanchaca de Bolivia, a Bolivian-controlled company, supplied nearly all of the remainder.

Most of the silver is obtained in Chile as a by-product from copper. During 1930, the output was distributed among the copper companies as follows: the Andes Copper Mining Co., a subsidiary of the Anaconda Copper Mining Co., 35 per cent, the Poderosa Mining Co., Ltd., and the Soc. des Mines de Cuivre de Naltagua about 11 per cent each, while of the silver companies, Cia Minas Beneficiadora de Taltal produced about 25 per cent and Soc. Beneficiadora de Plata de Condoriaco about 15 per cent. The production of silver in Colombia is of little consequence. The Frontino and Bolivia Gold Mining Co., a British firm, is an important producer, although the Soc. de Zancudo, a native company, has been the chief source of supply. The production of silver in Ecuador is not important and virtually all of it is yielded by the South American Development Co., an American gold-mining company.

Peru is the first country in South America in the production of silver. About three-fourths of the smelter production came from the Cerro de Pasco Copper Corp. and its subsidiary, Soc. Minera Backus y Johnston del Peru. The Northern Peru Mining and Smelting Co. accounted for most of the remainder. During 1929, about one-half of the silver was mined by the Cerro de Pasco and Backus y Johnston units, and 15 per cent by another American company, the Northern Peru Mining and Smelting Co. The Negociacion Minera Fernandini, a Peruvian company, produced about 13 per cent, while the Anglo-French Ticapampa Silver Mining Co. and the Lampa Mining Co., Ltd., added small amounts.

Sulphur.—Although Chile is one of the few countries where sulphur is produced on a commercial scale, its production is of little importance from a world viewpoint as it amounts to only 1 per cent or so of the world's total. Cia Azufrera y Minera del Pacifico, controlled by the British company, Tigon Mining and Finance Corp., Ltd., produced over one-half of the national output, in 1929, and the remainder came from small workings operated by Chileans. Small amounts of sulphur are produced in other countries for local use. No foreign capital is involved.

Tin.—Bolivia is the second tin-producing country of the world and yields about one-fourth of the supply. Almost two-thirds of the tin in Bolivia is produced from domestically owned mines and about 60 per cent from the Patiño properties, including Patiño Mines and Enterprises, Consolidated, in which considerable American capital is invested, Soc. Empresa de Estano de Araca, Cia Minera y Agricola Oploca de Bolivia and Empresa Minera de Huanuni. The firm of Hochschild, in company with other foreign groups, is interested in various mines and also is leading exporter of tin, handling about 15 per cent of the production. The principal distinctly American enterprises are the Caracoles Tin Co. of Bolivia, controlled by the Guggenheim interests, which produced about 8 per cent of the national output in 1929, and Cia General de Minas en Bolivia, which is operated by Easley and Inslee. Cie Aramayo des Mines en Bolivie, a Swiss-Bolivian company with considerable British and American capital, supplies about 5 per cent of the production. Among the British companies are Fabulosa Mines Consolidated, Duncan, Fox and Co., Philipp Bros. and Berenguela Tin Mines, Ltd., none of which, however, has a very large production. Soc. Estanifera de Morococala is an important producer and another Chilean firm, Monserrat, yielded a fair tonnage in 1929. Abelli y Cia, a domestic concern, and the French firm of Bebin Hermanos obtained tonnages of some consequence in that same year.

Titanium.—During 1929, a considerable amount of ilmenite was recovered from beach sands in Brazil. Two firms, Soc. Min. et Industrielles Franco-Bresilienne and Mauricio Isralson, operated at Esperito Santo, and one,

John Gordon, at Bahia. In the following year the production dropped to a small figure.

Tungsten.—A few tons of concentrates were produced in Argentina by a domestic firm in 1929 and in other years several small mines have been operated. During 1929, Bolivia was the second ranking country in the production of tungsten concentrate with about 10 per cent of the world's supply. About two-thirds of the concentrate was produced by the American firm of Easley and Inslee from their Conde Auqui mine, and most of the remainder by the Patiño interests at Kami. In 1930, the former was the only producer. Production in Peru is normally small.

Vanadium.—Peru is the world's leading producer of vanadium and all of it comes from the mines at Minasragra, owned by the Vanadium Corporation of America.

Zinc.—Cia Minera Aguilar, S.A., a subsidiary of the St. Joseph Lead Co., an American company, has developed a zinc-lead mine in the province of Jujuy but production has not commenced. Cia Huanchaca de Bolivia, a Chilean-French company, is the only producer of zinc in Bolivia. The Mathilde mine, and certain old dumps, are owned by M. Hochschild & Co. Practically all of the zinc in Peru is mined by two companies. Soc. Minera Backus y Johnston del Peru, a subsidiary of the Cerro de Pasco Copper Corporation, produces two-thirds of the tonnage and Negociacion Minera Fernandini, a Peruvian concern, produces the remainder. The Cerro de Pasco company has a large tonnage of lead-zinc ores that are, as yet, undeveloped.

Precious Stones.—The diamonds and carbonades of Brazil are produced by native diggers. Those of British Guiana are produced under the same conditions. At the time they were closed in 1931, the Chivor emerald mines, owned by the Colombian Emerald Development Corp., an American company, were the only emerald mines in operation in Colombia. Other properties are owned by the government. The pearling industry of Venezuela is carried on by local fleets of small fishing vessels.

SUMMARY

Any study of the ownership of the South American mineral resources brings out the important part played by American interests. The commodities produced from properties under their influence constitute an imposing list that includes all of the asphalt and bauxite, a considerable part of the coal, about 90 per cent of the copper, one-third of the gold, practically all of the iron ore, more than one-third of the lead, one-half of the manganese, over one-half of the petroleum, approximately one-half of the platinum, 70 per cent of the silver, only one-tenth of the tin, all of the tungsten and vanadium and two-thirds of the zinc. The most important items are copper and petroleum.

The minerals produced by local capital in the various South American countries make a list of about the same length; they are, most of the antimony, until recently all of the bismuth, most of the coal, a small share of the copper, about 15 per cent of the gold, two-thirds of the lead, one-half of the manganese, much of the nitrate, 5 per cent of the petroleum, about 30 per cent of the silver, one-half of the sulphur, two-thirds of the tin, and one-third of the zinc. Of this list, only the nitrates and tin are of major importance. While these lists are about equal in length, the chief products of the North American group, petroleum and copper, have about twice the value of nitrates and tin, the leading commodities in the South American list.

The British activity has been comparatively modest, for it is confined to about 8 per cent of the copper, 42 per cent of the gold, 5 per cent of the petroleum, one-half of the sulphur and 7 per cent of the tin. Obviously, from a world viewpoint, these tonnages are of little importance, but the petroleum production of a British-Dutch group, 35 per cent of the total for South America, is of considerable consequence. British capital has flowed into railways, lands, and other industries more freely than into mines in this region. French companies supplied small percentages of the copper, gold and tin.

In conclusion, the bulk of the productive mineral resources of South America are owned by American interests, most of the remainder by domestic capital, a small portion by British groups, and a substantial share of the petroleum by a British-Dutch combine, while the French have done little more than register their presence.

CHAPTER XIV

WHAT OF THE FUTURE?

ANY attempt at final answer to the questions raised in the first chapter evokes a lively sense of the fallibility of human judgment. It will be recalled that John Miers, in 1825, judged that there was little opportunity for copper mining enterprise in Chile, and that Sir William Crookes, in 1898, believed that Chilean nitrate resources would be exhausted by 1931. In both cases developments took place that were unforeseen and the predictions were falsified. The present authors can see no farther into the future than other men. But a sense of this difficulty is not sufficient excuse for failure to make such deductions from the available evidence as seem well founded. Before making such attempt it will be well to bring together in a concise summary the most significant among the data already presented in detail.

Five hundred years ago, Spaniards and Portuguese came to South America; seeking a short route to China, they found a new continent. The Spaniards saw little opportunity except to produce gold and silver and to baptise the natives. The Portuguese engaged in agricultural production, imposing less restriction than did the Spaniards, whose general policy was not to permit settlers in the new land to export anything that did not conform to their own conception of the best interests of Spain. In regions where the climate was suitable to African slaves they were imported to supplement the deficiency and inadequacies of native laborers, whose lot was usually no better than that of slaves, except in Argentina and Chile where they were too virile to submit. Such organized society as the Spaniards found in the portion of the country they occupied was based on agriculture. Mining was minor and incidental. They changed it but little, aside from intensifying the search for gold and silver.

Two centuries passed, marked by larger gold and silver production in Peru, Bolivia, Ecuador, Colombia and Venezuela, agricultural production in Brazil, and little change in Argentina and Chile. Then gold and diamonds were discovered in Brazil and for almost a century attention was focused upon them. Soon after 1800 came revolutions everywhere except in Brazil, whose independent status was peacefully attained. During the wars and in the period of uncertainty that followed, productive activity was generally reduced and took on a new cast, for the regions were now free to seek their self-interest instead of continuing to function to serve the interest of Spain. Argentina began the march that has increased its population ten times since 1850, and Chile began the copper production that was to lead into many developments. The rest of South America, including Brazil, changed but slowly. Gold discoveries in California and Australia about 1850 overshadowed gold production in South America, especially where it had been of minor importance. Copper production in the United States after 1880 caused a decline in the industry in Chile, but its effects were masked by developments in the nitrate industry, which for years thereafter was a "meal-ticket" for Chile. Copper developments in the United States reacted to cause a great renascence of copper production in Chile after 1900.

The World War, 1914–1918, stimulated mineral production throughout South America not only of copper, tin, manganese and nitrates, but that of minerals which had not previously been produced in important amounts, such as antimony, tungsten, chromite, and others. One or two effects of the war were inimical. It greatly hastened the inevitable development of synthetic nitrogen, and the world-wide search for manganese to which it gave rise proved unfavorable to production of the latter in Brazil, but the net results were stimulating. This was especially true in the case of petroleum. Venezuela, with no output before 1917, became for a time the second petroleum-producing country of the world, Peru's output increased ten times between 1910 and 1930, Argentina's yield multiplied by ten between 1916 and 1930, and Colombia's production rose from nothing in 1921 to 20 million barrels in 1930. South America's principal mineral commodity has successively been gold, silver, copper, nitrate, and now petroleum.

Production was always typically for export, as would be expected since the dominant interests of the continent were agricultural. Minerals were merely salable raw materials. Only within the past century has there been appreciable domestic consumption of minerals other than such common things as clay, stone, and lime. Petroleum in Argentina is the only example of large production of a mineral commodity wholly for domestic consumption.[1] The coal mined and used in Chile, Peru, Colombia, and Brazil is neither large in amount nor sufficiently good in quality to permit export. Its mining has, however, brought

[1] Much of Peru's exported petroleum is used in other South American countries.

about the best examples to be seen of a slowly developing balanced economy in those regions where such a development is possible.

In the attempt to visualize just what the mineral industry now amounts to in South America, a few consolidated figures for the year 1931 may be helpful despite the fact that they are largely estimates. Similar figures for more recent years, even where available, would present a distorted picture because of adverse world conditions. In 1931, then, the total number of workers in the mineral industries in eight of the countries for which estimates are possible was 305,900. Of these, 238,900 were in six countries where the total number gainfully employed was nearly 13,000,000, though the latter figure is not much more than a guess. Accepting it, however, less than 2 per cent of the workers were employed in the mineral industries. In Chile, where more complete data are available, 12 per cent were that year employed in the mines, about half the usual number. The detailed estimates are as follows:

Estimated Workers in Chile, 1931

Miners	38,400
Agricultural workers	136,000
Manufacturing laborers	72,250
Government employees	29,750
Railway workers	10,500
General civilian occupations	30,600
Miscellaneous and unclassified	334,400
	651,900

Among South American countries, Chile has an unusually well-balanced economy despite its large dependence on minerals for export. One of the striking and characteristic features of South American production and commerce is the large part the minerals play in export trade. Export of minerals is an important factor, not only from the point of view of creating exchange but in furnishing cargo for ocean shipping. Six of the countries shipped abroad 28,350,000 tons of mineral freight in the year taken for study. What this means in percentage of value of total exports is indicated in the table following:

EXPORTS OF MINERALS AND MINERAL PRODUCTS*

Country	Year	Total Exports	Exports of Minerals and Mineral Products	Per Cent of Total Exports
		Pesos	Pesos	
Argentina	1930	614,104,180	295,579	0.04
Chile	1930	1,328,122,967	1,110,393,207	83.60
Colombia	1930	112,708,549	37,250,527	33.05
		Milreis Paper	Milreis Paper	
Brazil	1931	3,398,222,000	58,849,000	1.73
		Soles	Soles	
Peru	1930	241,133,250	149,133,578	61.84
		Bolivars	Bolivars	
Venezuela	1930	762,494,233	661,263,755	86.72
		Bolivianos	Bolivianos	
Bolivia	1930	101,561,417	87,733,944	86.38
		Sucres	Sucres	
Ecuador	1930	80,646,539	23,774,913	29.48

* From a memorandum compiled by Palmer Pierce.

This is a substantial contribution to the buying power of the people of a continent. To make such exports possible has required investment of $1,607,000,000 in plant and facilities in only seven of the countries. Most of this money came in from outside. In five countries, where sufficient data exist to permit a rough estimate, it is believed that not more than $180,000,000 out of a total of $1,500,000,000 was derived from local sources. None of these figures should be taken as exact but they are of the right order of importance, and they indicate how heavily South America has been indebted to countries outside its borders for the development of its mines and treatment plants. Only a minor portion of this heavy investment has as yet been paid back. In one group of investments $110,000,000 now stands without return. Other instances have been cited in the preceding text. Of the gross returns from sales of the mineral it is sufficiently accurate to say that half is spent within the country of production for labor, local supplies, taxes, *etc.* The remainder must furnish the fund from which to meet bills for supplies bought abroad, which in practise means most of the machinery, for such foreign labor as is used, taxes in the country from which the money comes, the amortization of the capital, and bond interest; all this before there can be any profit for shareholders. In 1931, the three big American copper companies left $20,000,000 in Chile and took $6,000,000 out to meet the items enumerated above. Previously the proportions had been better balanced and in later years the return to shareholders was less, since expenditures on the ground, in the case of a mine, can never be reduced in the same ratio as production may have to be reduced.

In the table below, an attempt has been made to give a rough approximation of the net balance in value of exports and imports of the principal mineral commodities in South America. The approximation is rough indeed for the reported figures sometimes give quantities without values, as in Argentina, but even the quantities sometimes change abruptly from year to year. The official figures indicate that Argentina imported 224,000 barrels of cement in 1932 as compared to 2,713,000 barrels in 1929. Brazil did not report imports in 1931 of lubricating oil, though the actual value must be large. Imports of iron and steel are listed under such a variety of headings that the figure given is only a rough guess at a total from the detailed figures reported by a few countries. It does not include manufactures, such as automobiles, that are largely of iron, since there is no way of ascertaining the weight of iron involved, and to use the value would be misleading. This inability to record the metal in manufactures vitiates the balance for lead and zinc, as well as most other metals. The figures given in preceding chapters show the lead exported, in ore, concentrate and metal, from Bolivia and Peru, but they fail to indicate the lead content of imported articles of manufacture. It is most probable that South America on true net balance is an importer of both lead and zinc, while the net exports of copper and tin are smaller than the indicated gross figures by an unknown amount. Gold and silver are not included, since new metal is not distinguishable from bullion in the reports, and the net balance is small. The figures given below should therefore be taken only as a rough indication, for they are certainly inaccurate in detail.

NET BALANCE OF IMPORTS AND EXPORTS OF MINERAL COMMODITIES, SOUTH AMERICA, 1931, BY VALUE IN U. S. GOLD DOLLARS[2]

EXPORTS	
Crude Petroleum and fuel oil	$100,000,000
Copper	49,000,000
Nitrate	43,296,000
Tin	17,719,000
Platinum	1,192,000
Iodine	1,028,000
Manganese	450,000
IMPORTS	
Iron and steel, unmanufactured	$20,000,000
Coal	18,000,000
Gasoline	16,000,000
Kerosene	7,000,000
Lubricating oil	8,000,000 (?)
Portland cement	3,000,000

[2] Figures compiled from Foreign Commerce Yearbook, 1933.

Incomplete figures for 1932 indicate a greater decline in exports than in imports; nitrate declined almost $40,000,000, copper by over $30,000,000, and tin by nearly $10,000,000. Were it not for exports of crude petroleum, South America would possibly in 1932 have been a net importer of mineral commodities, especially if heavy chemicals, such as soda ash and dynamite, were taken into account; it unquestionably would have been if the mineral content of imported manufactures could be assessed.

So long as the external trade of South American countries consisted almost wholly of exporting natural resources, particularly during the sixteenth, seventeenth, and eighteenth centuries, when mineral exports were almost entirely gold and silver bullion, miners had little occasion to worry about foreign-exchange problems, but as soon as the standard of living in those countries started to rise, creating a demand for foreign goods which had to be imported, it was inevitable, human nature being what it is, that sooner or later there would come a time when the value of exported products and services would be insufficient to satisfy the desires of the population. That situation creates foreign-exchange problems. One common solution, with nations as with individuals, has been to borrow money to pay for the excess cost of imports over exports, and that was done by South American countries on a fairly large scale during the nineteenth century and during the first quarter of the twentieth century. When this means of balancing payments was cut off about 1928, South American countries could satisfy their wants only so long as the price of the products they exported (practically all minerals or agricultural products; *i.e.*, raw materials) was sufficient to pay for their imports. When the prices of these raw materials collapsed, a serious foreign-exchange problem was inevitable so long as the populations involved insisted on importing articles they had been used to. In practice it has proved impracticable to allow a "natural" adjustment between interests who wish to import and interests who had foreign exchange available from exports. When chaos threatened, the States stepped in and made the decision as to how the insufficient foreign exchange available should be allocated. This step was taken in most South American countries toward the end of 1931 after Great Britain had abandoned the gold standard, and exchange controls have, since that time, been a major factor in operating mineral enterprises in South America.

This has been particularly true because the mineral enterprises, together with the agricultural, have been practically the sole sources of foreign exchange available. It was predictable that the operation of exchange control would be particularly embarrassing to these enterprises, and ultimate relief can be expected only when the price of exported products in each country rises sufficiently to take care of the import obligations, including capital and interest charges, presumably on a reduced basis. This is the fundamental situation behind the incredibly complicated maze of quotas, tariffs, subsidies, and blocked currencies which make the operation of any foreign enterprise so difficult today.

A concrete instance will perhaps best illustrate how this factor currently operates. In the annual report of Patiño Mines and Enterprises Cons., Ltd., for the year ended December 31, 1933, it is stated that the company was obliged to deliver to the Banco Central de Bolivia a specified proportion of their drafts on London, the amount required in March, 1933, being 52 per cent. For the drafts so delivered, they received Bolivian currency which was in excess of their requirements for operating expenses, so that by the end of the year they had accumulated cash in Bolivia amounting to 2,020,142 Bs. (£110,000). As a result of this blocking of exchange the company had reduced its output by 100 tons of tin per month below its allotment under the international cartel, in order to avoid accumulating too large a cash balance.

In a similar way, the petroleum companies selling refined products of petroleum in Brazil were unable to remit more than a fraction of the proceeds of their sale to the country of their origin, and for a time experienced much difficulty in carrying on their business. It is unnecessary to multiply other instances, since these two diverse ones will make the point clear that a new hazard has been added to the already sufficiently great ones of mineral enterprise. While the averred purpose of exchange control is to stabilize it, fluctuations in exchange in recent years have been wide. It is not sufficient for a mineral enterprise to calculate production costs, transportation and other charges, and forecast the probable selling price of its product in order to form a judgment as to whether the enterprise is likely, or not, to be profitable, but exchange must also be taken into account, since its range of fluctuation may be wide enough to turn what would otherwise be a profitable enterprise into a losing one.

Various expedients have been suggested to meet the first aspect of this problem, inability to return to country of origin capital invested in South American countries. One possibility would be for a prospective investor to borrow within the country the money needed to finance the enterprise, but it does not seem likely that South American banks would be willing to extend long-term loans that might be repaid out of operating profits, and no operator would be willing to take the risk involved in heavy short-term loans or ordinary bank credits. Another suggestion is that the profits of one enterprise be used to develop another within the same country, but this involves the difficulty of finding another deposit that seems worthy of exploitation, or the alternative difficulty of venturing into a field, such as railroad building and operation, in which the company has not sufficient background of experience to permit it to proceed safely. In any case, this method would merely postpone to a future date the returning of their capital to investors, with no guarantee that future facilities for so doing will be any better than they are at present, although world-wide price recovery should eventually solve the problem.

While not directly related to exchange, the emergency taxes that have been imposed on enterprises in South American countries during the past few years have their origin in the general financial situation of the countries, and are therefore definitely related to it. They add a third hazard to the financial success of a mining venture. To discuss them and the problems of exchange in more detail would take us too far afield from our main purpose, but they must at least be mentioned in attempting a forecast of the future of mineral enterprise in South America.

It is clear from what has been already said that a belief that South America is a vast reservoir of untouched mineral wealth is wholly illusory and that the cost of producing such as exists has so far left but moderate returns to those responsible for it. Only in the case of possible future discovery of petroleum in large quantities in the immense area lying east of the Andes is there any basis for hope of big returns, and even if such a vision should come true, the capital investment required would be so great that most of the returns would be needed to finance the transportation facilities and employment, which in turn would lead to general economic development rather than heavy profits to the capitalists. The immense iron resources of Brazil and the less impressive ones of Chile and Venezuela will have to be mainly produced for export. It is at least doubtful whether they will be able to bear more than a small export tax, certainly for a long time in the future. The day of Chile's nitrate monopoly seems definitely over, and the future for copper production, except for the lowest cost companies, is clouded. Tin is of importance only in Bolivia and production there will hereafter be limited to a fraction of a world quota.

Of the three big mineral industries of the world, coal, petroleum, and iron and steel production, South America has no hope of becoming of importance in the first and for the third must be mainly content to furnish ores for reduction elsewhere. Only as regards petroleum can the continent be regarded as a "land of the future" and that future may measure nothing more than retaining a degree of importance which it has already attained. Of the three minerals of next importance, copper, lead and zinc, it has no certain future as regards the latter two and no large hope of exceeding its past rank as a producer of copper. Gold, as has been repeatedly said, was of importance in a day when world production was small; even if its production now rose to its former highest level it would still be only of minor significance in a world accustomed to greatly increased magnitudes. Silver's future is coupled in South America with that of copper and tin, while silver at its best is a metal whose future price is likely to fall unless it is politically pegged.

Turning to the question as to what mineral production has done for South America, it is clear that it has furnished a job for a small fraction of its people. So far as the natives were concerned, they did not want the job, nor did it result in any betterment of their way of living in the Colonial period. It has been left to the large corporations during recent years to take those steps which have brought about better health, better diet, better housing, more entertainment, and to provide a practical education for the worker which increases his earning power and raises his standards of living in ways that have been instanced in the preceding chapters. It may well be questioned how far this can proceed in regions where the aborigine forms a large proportion of the population. In the United States an almost equal length of contact with the white man has had but little economic influence on the Indians except in the southwest, where the introduction of sheep-raising and wool spinning has somewhat improved their economic status and led to an increase in their number. But it seems significant that after a century and a half of experience with Indians as wards of the government our own Indian Office has recently announced a retreat from industrial schools and a policy of "civilizing" the Indian to one which aims to produce a renascence of their native culture. The lot of a Quichua or an Aymara today is somewhat better than it was in the sixteenth and seventeenth centuries, but not much, if any, better than it was before the Spaniards arrived, and it is doubtful if it ever will be. They had then developed a way of living that seemed to suit them and their environment and it is possible that nothing better can be found for them as a whole. It has, however, already been pointed out that because of the power factor in mineral production a relatively small portion of a total population produces a relatively large tonnage, and even value, of mineral. It is also to be remembered that in a dominantly agricultural economy, even in these days, the amount of mineral called for is limited. The minerals furnish the materials for the heavy industries, for construction, and so for the larger life. If it accords with the desire of the people they may be used as raw materials to trade for these luxuries and the main mass of the people remain as before on the farms. That in fact is exactly what South America, by and large, is now doing.

Whether or not the actual desires of the people will change in any large way depends on many factors, one of which is clearly the amount of infusion of blood from other races with a different inheritance. The policy of intermarriage between whites, natives, and Negroes that has prevailed in South America is thus of concern because of its economic implications. A cattle-raiser understands what results he will get by crossing "scrub" cattle with pedigreed sires, it will be a "grade" herd. For beef-raising this is better economics than trying to maintain a pure-bred herd, but it is not known whether with human beings a similar economic result will be attained. It is now clearly understood that mass production can only develop *pari passu* with mass consumption and the "grade" man is only a grade consumer, since what he will want and work for is a result of mental rather than physical characteristics. If the countries of South America are to attain a balanced economy on a high level for the average man, it will be because the average man wants more things and is willing to work to attain that end. If the will to the end is lacking, it will never be attained.

There is no question that some of the people in each country have a keen appreciation of the benefits of a higher standard of living, nor can it be doubted that the mining enterprises have served as a center from which the influence of such desires has spread. Mines are necessarily where ore occurs, and as a rule the country around a mine is good for little else. To be worked at all in South America, a mine must be large and a mining company a strong organization. Accordingly, it dominates its community and to an extent beyond that of other enterprises sets local standards of living. No one who has visited many large mines in South America will doubt that the standard of well-being in the communities where mining is undertaken is materially raised.

Mine managers long since learned that it is not the rate of pay per day which is of first concern, but the cost per unit of work, and that the road to low costs is through increased efficiency of the workers. Therefore they put more and more mechanical power back of each worker, steadily improve his housing and his food, and encourage the growth of his desires in order to not only make him a better physical working machine, but to increase his incentive to work. It has been shown over and over again in South America, as elsewhere, that this is a thoroughly practical road to success and it has also been demonstrated that with skillful management, pushing neither too fast nor too slow, it is possible with the labor actually present not only to attract the better workers but to make better and more willing workers out of the mass. Some years ago, the South American Development Co. at Zaruma, in Ecuador, introduced free meals for its workers, with marked benefit both to them and to the company. It was necessary to supply the meals, plenty of hot cooked food, at eating places maintained by the company rather than to issue rations. The latter would merely have attracted more dependent relatives from the mountains and the actual workers would have been no better fed than before. To have increased wages instead of giving the meals would, under the conditions there at the time, merely have enriched the gamblers. So management as well as mere will is necessary to produce results.

That even excellent laborers can, when they want, increase their output is also well known. The miners in the nitrate fields have long had an exceptional reputation for the amount of caliche they delivered per man-day under an ingenious system of piece work. Not only were surprisingly low costs achieved but the men and their families were better housed, better fed, and had more amusements than before they came to the mines. The standard of living has been steadily rising, and the men have through the years also steadily increased their deposits in the government savings bank, to the ultimate end of purchasing a bit of land in the central valley and becoming themselves landowners rather than peons. Even here a little extra incentive resulted in more *caliche* on the *cancha*, as was proven over and over again by the larger payrolls it was necessary to meet each fall on paydays before the annual celebration on the Eighteenth of September. Like the small American boy who becomes interested in nickels when the circus is billed to come to a town, the *particulare* in the nitrate fields puts extra energy into his work so as to be the better prepared for the national festival. These are small matters, perhaps, but they show that Indian and mestizo labor is responsive to many of the same incentives as are workmen elsewhere, and perhaps indicate that the will to work can by proper methods be aroused to an extent not generally appreciated. That such laborers can be made into skillful workmen when their interest is aroused is proved by many circumstances in all the countries. One instrument which is operating to bring this about is the motion picture. Crude as many of the films are, and lurid and untrue to life as are many shown in the camps and small towns, they do show, even to those who cannot read the titles, workmen of other lands living under conditions far superior to their own. That they lead to a desire to emulate is not always the happy result which might be anticipated. It is of record, for example, that the first train robbery in a certain foreign land followed the local showing of an American film. The films certainly are educational.

In mass movements the speed of a convoy is that of its slowest ship and in a country where labor has been accustomed only to the simple tools and small units of operation characteristic of agriculture and cattle raising, immediate change into highly industrialized industry with big integrated units and liberal use of mechanical power is not to be expected. This is even the more true in South America where the older social organization was aristocratic. Large land holdings, patron and peon, have been the rule, and in the long run the wealth of the few who were patrons led to a virtual monopoly for that class of the opportunity for advanced schooling. It became distinctly "the thing" to be a land owner, and aside from that status unless one could be a doctor, a lawyer, or a government official, an individual lost what the Far Easterners call "face." These particular professions have become crowded and mechanical power has not yet operated to build up a large middle class. To a degree surprising to a visitor, the actual trade and commerce, even down to small shops, is in many of the countries in the hands of foreigners, another characteristic of lands wedded to the aristocratic form of social organization. Human nature, however, has much in common in all lands. In one South American country the Ford Motor Co. opened an assembly plant in the days when

Mr. Ford decreed a minimum wage of $5 per day, without much thought of differences in wage standards in different countries. Translated into local pesos this wage was so large as immediately to dignify labor beyond all past local records. A nephew of the President, sons of influential Senators, and other young men of the ruling class scrambled for jobs in the plant and for once it became "the thing" to be a workman as well as a landowner. How far these new influences will operate no one can rightly say in advance, but it is significant that they are operating and that the mineral industries are playing a large part in their introduction.

On the material side the mineral industries share with agriculture the credit for having provided South America with that degree of port facilities, inland transportation, and urban development which it has attained. The relative part played by each varies. In Argentina, development has been almost wholly due to agriculture and mineral production there has been in response to needs created by an improved economic status, rather than a cause of it. In Brazil, the situation has been much the same, but in Chile, mineral enterprise was a cause rather than an accompaniment of advancement. In Peru, mineral industry played an even larger part as the cause of material progress, while in Bolivia, it has been almost the sole cause. Everywhere the entrepreneur has turned his hand to whatever promised the most advantage in exchange of goods, though it is probably not far wrong to count the rôle of gentleman-farmer as representing most nearly the Latin-American concept of the good life, and next to that the rôle of successful politician. It has not been merely greater willingness to risk capital and a better knowledge of technology that has made the foreigner the leading spirit in mineral enterprise; except to

Chileans, and to a less extent to Peruvians, the rôle of mineral producer, other than of gold and silver, has no great appeal. Where mineral enterprise has largely benefited South America it has most often resulted from the initiative of foreigners seeking an opportunity to make a profit from productive enterprise. It has not been a missionary enterprise based on an altruistic wish to benefit the natives, nor has it been a predatory "exploiting" of the resources of the country; it has simply been *laissez-faire* capitalism functioning in the way the physiocrats believed that it should.

Such investments require not only the courage that Clarence Wood showed when, his principals having "turned down" the gold mine he had found for them in Peru (but at the end of a forty-mile muleback trail connecting with a 100-mile road to a 14,000-foot elevation railway station), he took it up in his own name and found the capital to develop it, but also long and heart-breaking waiting for returns. It was fifteen years from the organization of the Cerro de Pasco to the first dividend, and meanwhile $25,000,000 had been invested, not of money collected from the public, but the private funds of Mr. Haggin and his associates. Anaconda had been investing money in the Poterillos property for ten years, and had put in $20,000,000 before public financing was done. The original promoters of Braden thought they had a million-dollar property, but $14,000,000 was put into it before it was fully developed; the original promoters surrendered their control of the property because the money required was beyond their financial depth. Such enterprises were no random quest for a gilded man, but the courageous venturing of large amounts of capital in enterprises that had been carefully studied and judged worthy. If it were possible to set alongside of these examples of success a record of those who put money in and failed to secure a return, the balance might well be on the other side. There is a lot of money lying unproductive in oil ventures in Colombia. In Peru, one exploration syndicate spent over a million and a half under the best engineers trying to find a mine and withdrew without finding any which in their judgment warranted development. Such instances might be multiplied, but these will serve to make clear that successful mines must pay for many failures or mining would stop.

It is often claimed that South America offers to the United States a rich market for the sale of mining machinery, and other equipment for its mineral enterprises. This is a superficial view, for it overlooks the fact that the capital would be raised in the United States. If the sale had been made to a property in the United States and a market could be found for the output, the equipment would have remained under our control and would have provided employment for our citizens. A good many "sales" of equipment built in the United States, to South American enterprises differ but little from the "loans" the United States made to European governments to finance their purchase of war equipment in our country. The net result in both cases was that the United States provided large quantities of material things for use in a foreign land; the difference in the ways in which individual Americans were involved in the transaction should not obscure a clear vision of the transactions as a whole. Mining investments in South America did not even result in purchase of American equipment in many instances, for many cases could be cited where the capital raised in the United States was used to purchase German-built Diesel engines and other European products. Probably the chief attraction of the South American market to builders of American machinery resides in the belief in the great richness of South America that has made it relatively easy to promote and finance ventures there.

During its first three centuries South America was merely an exporter of minerals, since its needs for more than a few simple tools of iron were but slight. The past century has seen a great growth in mineral imports, beginning with rails and locomotives and developing to where it embraces an immense variety of articles of manufacture that are essentially mineral in their make-up. It is said that an automobile employs more than sixty varieties of metal in its construction and that statement throws a flood of light on why the production of more than ore and crude metal is slow in developing in South America. The production of each variety in the standard of quality and with the physical characteristics demanded requires not only equipment and skill but a market for considerable amounts of that variety. Even when the amounts required are considerable and the quality a standard one, as in the case of iron pipe, a variety of sizes are needed and the methods of manufacture are such that what is needed will bear the cost of transportation from established plants unless some artificial means of stimulating local production is resorted to. The discussion in the chapters on Argentina and Brazil of the problems of domestic production *versus* reliance on imports applies to the whole of South America. With the continent able to supply its requirements for petroleum from within its borders, trade routes and transportation facilities still make it more economic to draw at least part of its supply from other lands while simultaneously exporting. Geographical rather than political subdivisions are the natural factors in trade. Political considerations lead to attempts to circumvent geography, but they can never be entirely successful.

The conclusion from this is that South America is destined for some considerable time in the future to remain in its past rôle of exporter of minerals either in the raw or but simply processed state. Meanwhile, its needs for minerals in manufactured form are steadily increasing. Venezuela and Peru can produce more refined petroleum than they need, Argentina lacks crude petroleum rather than the facilities to put it in form for use. But no South American country has more than rudimentary facilities for the manufacture of iron and steel, or for the fabrication of other industrial metals. What might be described as the density of demand for these is still too slight to permit more than slow and simple development, and no such rapid industrialization as took place in Pennsylvania and Ohio can be expected anywhere in South America. One reason is the lack of an abundant supply of cheap fuel, but the underlying one is that both the natural resources of the continent and the habit of thought of its peoples favor making agriculture the dominant industry. Manufacturing will continue to increase and both it and agriculture will call for an increasing amount of mineral product. Where local conditions favor, the local minerals will be worked up, as in the case of the small cement mills now existing where transportation costs protect them from competition. Under the impulse of national aspirations and national planning, local industries may be expected to be stimulated and protected, but only as in Argentina, where a large demand for petroleum exists and the physiography permits cheap distribution over a considerable industry, is it likely that such industries unless immoderately subsidized can stand the competition of the mass-production units in the more favored industrial countries. It will continue in general to be more profitable to sell the surplus of mineral raw materials and purchase finished or semifinished goods than to manufacture on any practicable scale. No large home-built fleet need be expected to come out of the South to disturb the world. Instead, the major influence of the South American countries seems certain to be on the side of peace in world affairs, to keep open the channels of trade and make firm the bonds of good will.

South America is, indeed, a continent with a future, but not a spectacular one in industry. Here and there an adventurer will find a gold mine; from many countries a surplus of minerals will be shipped as raw materials; petroleum production may be expected to increase and important new fields may be found; in individual areas mineral fabrication will develop as a part of manufacturing to meet local needs; but on the whole the world may expect the development to be slow and along existing lines. Industrialization in much of the world through the past half-century has proceeded at a dizzy pace. There are those who question whether it has not gone too fast and too far for man to accommodate himself to it. Possibly the conservative forces in the countries to the south of us have a value to mankind we have not entirely appreciated.

INDEX

INDEX

INDEX

INDEX

INDEX

INDEX

INDEX

INDEX

INDEX

AMERICAN BUSINESS ABROAD

Origins and Development of the Multinational Corporation

An Arno Press Collection

Abrahams, Paul Philip. *The Foreign Expansion of American Finance and its Relationship to the Foreign Economic Policies of the United States, 1907-1921.* 1976

Adams, Frederick Upham. *Conquest of the Tropics:* The Story of the Creative Enterprises Conducted by the United Fruit Company. 1914

Arnold, Dean Alexander. *American Economic Enterprises in Korea, 1895-1939.* 1976

Bain, H. Foster and Thomas Thornton Read. *Ores and Industry in South America.* 1934

Brewster, Kingman, Jr. *Antitrust and American Business Abroad.* 1958

Callis, Helmut G. *Foreign Capital in Southeast Asia.* 1942

Crowther, Samuel. *The Romance and Rise of the American Tropics.* 1929

Davids, Jules. *American Political and Economic Penetration of Mexico, 1877-1920.* 1976

Davies, Robert Bruce. *Peacefully Working to Conquer the World:* Singer Sewing Machines in Foreign Markets, 1854-1920. 1976

de la Torre, Jose R., Jr. *Exports of Manufactured Goods from Developing Countries.* 1976

Dunn, Robert W. *American Foreign Investments.* 1926

Dunning, John H. *American Investment in British Manufacturing Industry.* 1958

Edelberg, Guillermo S. *The Procurement Practices of the Mexican Affiliates of Selected United States Automobile Firms.* 1976

Edwards, Corwin. *Economic and Political Aspects of International Cartels.* 1944

Elliott, William Yandell, Elizabeth S. May, J.W.F. Rowe, Alex Skelton, Donald H. Wallace. *International Control in the Non-Ferrous Metals.* 1937

Estimates of United States Direct Foreign Investment, 1929-1943 and 1947. 1976

Eysenbach, Mary Locke. *American Manufactured Exports, 1879-1914.* 1976

Gates, Theodore R., assisted by Fabian Linden. *Production Costs Here and Abroad.* 1958

Gordon, Wendell C. *The Expropriation of Foreign-Owned Property in Mexico.* 1941

Hufbauer, G. C. and F. M. Adler. *Overseas Manufacturing Investment and the Balance of Payments.* 1968

Lewis, Cleona, assisted by Karl T. Schlotterbeck. *America's Stake in International Investments.* 1938

McKenzie, F[red] A. *The American Invaders.* 1902

Moore, John Robert. *The Impact of Foreign Direct Investment on an Underdeveloped Economy: The Venezuelan* Case. 1976

National Planning Association. *The Creole Petroleum Corporation in Venezuela.* 1955

National Planning Association. *The Firestone Operations in Liberia.* 1956

National Planning Association. *The General Electric Company in Brazil.* 1961

National Planning Association. *Stanvac in Indonesia.* 1957

National Planning Association. *The United Fruit Company in Latin America.* 1958

Nordyke, James W. *International Finance and New York.* 1976

O'Connor, Harvey. *The Guggenheims.* 1937

Overlach, T[heodore] W. *Foreign Financial Control in China.* 1919

Pamphlets on American Business Abroad. 1976

Phelps, Clyde William. *The Foreign Expansion of American Banks.* 1927

Porter, Robert P. *Industrial Cuba.* 1899

Queen, George Sherman. *The United States and the Material Advance in Russia, 1881-1906.* 1976

Rippy, J. Fred. *The Capitalists and Colombia.* 1931

Southard, Frank A., Jr. *American Industry in Europe.* 1931

Staley, Eugene. *Raw Materials in Peace and War.* 1937

Statistics on American Business Abroad, 1950-1975. 1976

Stern, Siegfried. *The United States in International Banking.* 1952

U.S. Congress. House of Representatives. Committee on Foreign Affairs. *The Overseas Private Investment Corporation.* 1973

U.S. Congress. Senate. Special Committee Investigating Petroleum Resources. *American Petroleum Interests in Foreign Countries.* 1946

U.S. Dept. of Commerce. Office of Business Economics. *U.S. Business Investments in Foreign Countries.* 1960

U.S. Dept. of Commerce. Office of Business Economics. *U.S. Investments in the Latin American Economy.* [1957]

U.S. Dept. of Commerce and Labor. *Report of the Commissioner of Corporations on the Petroleum Industry:* Part III, Foreign Trade. 1909

U.S. Federal Trade Commission. *The International Petroleum Cartel.* 1952

Vanderlip, Frank A. *The American "Commercial Invasion" of Europe.* 1902

Winkler, Max. *Foreign Bonds, an Autopsy:* A Study of Defaults and Repudiations of Government Obligations. 1933

Yeoman, Wayne A. *Selection of Production Processes for the Manufacturing Subsidiaries of U.S.-Based Multinational Corporations.* 1976

Yudin, Elinor Barry. *Human Capital Migration, Direct Investment and the Transfer of Technology:* An Examination of Americans Privately Employed Overseas. 1976